Evolutionary Studies

Series Editor
Naruya Saitou, National Institute of Genetics
Mishima, Japan

Everything is history, starting from the Big Bang or the origin of the universe to the present time. This historical nature of the universe is clear if we look at evolution of organisms. Evolution is one of most basic features of life which appeared on Earth more than 3.7 billion years ago. Considering the importance of evolution in biology, we are inaugurating this series. Any aspect of evolutionary studies on any kind of organism is a potential target of the series. Life started at the molecular level, thus molecular evolution is one important area in the series, but non-molecular studies are also within its scope, especially those studies on evolution of multicellular organisms. Evolutionary phenomena covered by the series include the origin of life, fossils in general, Earth–life interaction, evolution of prokaryotes and eukaryotes, viral and protist evolution, the emergence of multicellular organisms, phenotypic and genomic diversity of certain organism groups, and more. Theoretical studies on evolution are also covered within the spectrum of this new series.

Shigeki Mitaku • Ryusuke Sawada

Evolution Seen from the Phase Diagram of Life

Springer

Shigeki Mitaku
Nagoya University
Kokubunji, Tokyo, Japan

Ryusuke Sawada
Department of Pharmacology, Graduate School of Medicine, Dentistry and Pharmaceutical Sciences, Okayama University, Okayama, Japan

ISSN 2509-484X ISSN 2509-4858 (electronic)
Evolutionary Studies
ISBN 978-981-97-0059-2 ISBN 978-981-97-0060-8 (eBook)
https://doi.org/10.1007/978-981-97-0060-8

The English manuscript was created with the help of artificial intelligence. A subsequent human revision was done primarily in terms of content.

This Springer imprint is published by the registered company Springer Nature Singapore Pte Ltd.
The registered company address is: 152 Beach Road, #21-01/04 Gateway East, Singapore 189721, Singapore

Preface

It is a great challenge to reveal the evolution of living organisms through a physical approach, and I (Mitaku) have been working on this problem for nearly 40 years. This book, *Evolution Seen from the Phase Diagram of Life*, is the result of nearly 40 years of research.

It was in the mid-1980s that I first aspired to understand the physics of living organisms, including their macroscopic aspects. At that time (and even still), the usual approach to the study of living organisms was to take a bottom-up approach, building from individual molecules to the whole. My idea, on the other hand, was that a top-down approach was needed to understand the macroscopic aspects of living organisms in physical terms. In other words, we had to boldly coarse-grain the organism. The result of that research was SOSUI, a highly accurate membrane protein prediction system using only physical parameters. This was published shortly before the year 2000, but it was still based on the idea of coarse-graining at the amino acid sequence level and did not reach the macroscopic genomic level of the organism.

In approximately 2000, with the Human Genome Project, it became possible to analyze the genome sequences of various organisms. At the same time, the physical study of the macroscopic aspects of organisms became possible. The first physical approach to our whole genome study was an analysis of the percentage of membrane proteins in the genome, and the results showed that the percentage of membrane proteins was constant (1/4) in all genomes. This led to a macroscopic order of the organism that could not be seen from the study of individual proteins.

This discovery led to a very important realization. In living organisms, as in matter, order can be formed from random processes. Just as the distribution of random molecular motions in matter is constant in equilibrium, the distribution of membrane proteins is also constant in organisms when random mutations are in equilibrium. We then moved on to the study of mutation simulation at the genome level, conservation laws for nucleotide composition at codon letter positions, and the phase diagram of life in compositional space. From this stage onward, another author (Sawada) was involved in the research. The research progressed comparatively smoothly, including evolution based on the phase diagram of life, discrimination of species and genera based on genome sequences, and analysis of viral genomes. The only problem was that it was very difficult to write the background of the research in a short paper. Therefore, we decided to present our research to the world in the form of a monograph.

However, this book does not explain many important facts about the biological sciences. For example, the codon usages are exactly the same as the nucleotide composition for each letter position of the codon we have used. The only difference is whether the argument is based on amino acids or DNA. We have taken the latter view because we believe the latter is more essential for discussing the macroscopic order of organisms. In addition, we hardly mentioned the study of the vast number of individual genes obtained by genome analysis. This is because, in discussing the macroscopic order of biological genomes, it is better to treat information on the active sites of individual genes as noise.

The book is divided into four parts. Part I asks big questions about living organisms. There, I stated that understanding organisms requires two approaches, bottom-up and top-down, and that they must be integrated. In Part II, I summarized the results to date of bottom-up research on organisms. There, I touched as much as possible the connections with Parts III and IV, which discussed top-down research. In Part III, we introduced the physical analysis of whole genome sequences along with the studies we have conducted. We showed that the nucleotide composition at each codon letter position is the parameter for the phase diagram of life. In Part IV, we showed that the species can be defined by the position in the phase diagram of life. Finally, we concluded the book by presenting our analysis of viral genomes and enumerating the remaining problems.

Many people helped me in the course of my research. I am very grateful to my collaborators, including the students I worked with. In particular, I am deeply indebted to Dr. Takatsugu Hirokawa for his great contribution to the development of the membrane protein prediction system. I am also grateful to Dr. Ryusuke Sawada (coauthor of this document) for advancing many important research projects. Dr. Koji Okano, who is well known in the fields of liquid crystal physics and polymer physics, taught me the concept of coarse-graining. I also learned the big ideas of research from Dr. Akiyoshi Wada, who triggered the Human Genome Project and led the establishment of the Human Genome Analysis Center in Japan. I am very grateful to Dr. Naruya Saitou for his great help in the process of publishing this book. I also thank Dr. Nobuyuki Uchikoga for his comments on the manuscript. Discussions with various people, both positive and negative, were also very helpful.

Finally, I would like to express my gratitude to all involved.

Kokubunji, Tokyo, Japan
September 2023

Shigeki Mitaku

Contents

Part I
The Big Question About Living Things

Chapter 1
Organisms Viewed from Real Space and Sequence Space

Keywords Bottom-up approach · Top-down approach · Human genome project · Codon table · Law of increasing entropy

Biology has a very long history. In the process, much research has been conducted on the classification of organisms, the discovery of microorganisms, and the structural analysis of biological macromolecules. Furthermore, in the twenty-first century, it has become possible to analyze the genome sequences of all living organisms. Since the genome sequence contains all the genes of an organism, it is now possible to select research targets from all genes and proteins. The impact of this genome sequence analysis has been significant and has greatly advanced the bottom-up approach to biology. In Chap. 1, we discuss the history of genome analysis and the impact of genome sequences on biology. We also discuss how genome sequencing has enabled not only the traditional bottom-up approach but also the top-down approach in biological sciences. This is because whole-genome sequences contain significant implications that have yet to be elucidated.

1.1 Conversations Between Biologists and Physicists

I (one of the authors: Mitaku) once witnessed a discussion between a biologist and a physicist. The physicist asked the biologist the following question.

> It seems that organisms evolved to become what they are today. Why do they become more complex over time? In physics, matter does not naturally become more complex because of the law of increasing entropy. This is true in a system without energy or matter inputs and outputs. Therefore, it is theoretically possible for an organism with energy and matter inputs and outputs to become more complex naturally. However, it is not clear why they change unilaterally to more complex.

Of course, there were objections from biologists. However, before introducing those objections, it is necessary to explain the minimum knowledge about life in the early 1970s, when this discussion took place. Basic knowledge about the genetic phenomena of living organisms is already available. That is, the fact that the genetic

S. Mitaku, R. Sawada, *Evolution Seen from the Phase Diagram of Life*, Evolutionary Studies, https://doi.org/10.1007/978-981-97-0060-8_1

code is written on the DNA molecule in the cell (Avery et al. 1944), the central dogma that converts a DNA sequence into an amino acid sequence via an RNA sequence, and the codon table that converts three nucleotides into one amino acid were already known. However, at the time of the above discussion, it was not conceivable to analyze the entire DNA sequence, the entire genome sequence, which designs the individual organism.

A few more details about the knowledge at that time need to be explained (Sadava et al. 2008). First, it was revealed by O. Avery and others in 1944 that the genetic code was written in the DNA base sequence (Avery et al. 1944). Then, in 1953, the double helix structure of DNA was discovered by Watson and Crick, and the mechanism by which genetic information is transmitted to the next generation was elucidated (Watson and Crick 1953). Figure 1.1 shows a model of DNA, in which two bases are placed inside the main chain of sugar and phosphate in DNA. Then, when adenine (A) binds to thymine (T) and guanine (G) binds to cytosine (C) in a complementary manner, the distances between the ends of the base pairs match exactly. As a result, a beautiful double helix structure can be formed, regardless of how the bases are arranged. Thus, if one strand is used as a template to synthesize the other strand, the next generation of DNA sequences can be obtained, and genetic information can be transmitted. In other words, DNA is an extremely good information medium at the molecular level.

What is created based on the DNA sequence is the amino acid sequence of a protein. The amino acid sequence is then folded together to form a protein that performs a biological function. Proteins are made of amino acids and have a unique molecular structure that is completely different from that of DNA. Therefore, the first question is how the amino acid sequence is formed from the DNA sequence. This flow of information is called central dogma. As shown in Fig. 1.2, the DNA sequence is first transcribed into an RNA sequence. RNA is a very similar molecule

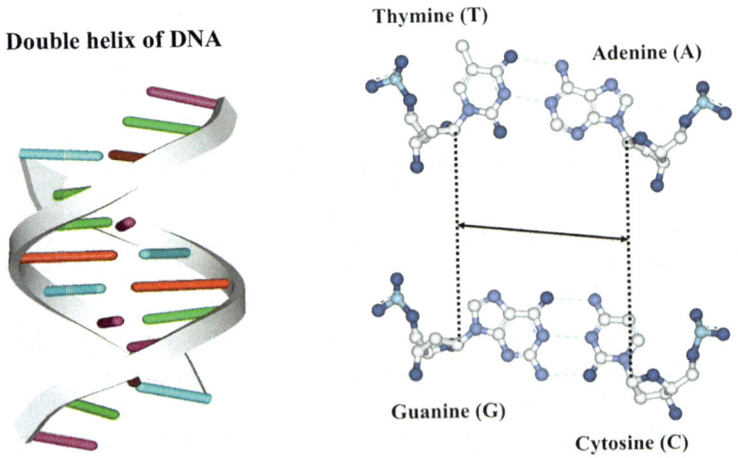

Fig. 1.1 Model of DNA double helix and base pairing. From Mitaku (2015)

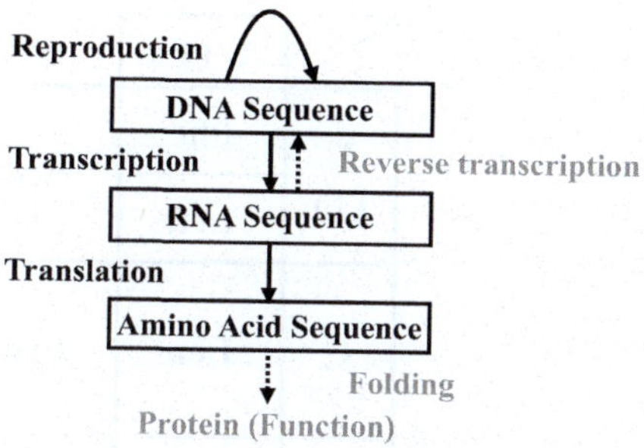

Fig. 1.2 Genetic information flow (central dogma). From Mitaku (2015)

to DNA, with the main chain portion being deoxyribose in DNA but ribose in RNA. However, the thymine base in DNA is uracil in RNA. The synthesis of DNA to RNA is performed by using the base sequence of DNA as a template and by using the complementarity of base pairs.

In contrast, RNA and amino acids have completely different molecular structures, so the actual process is very complex and is called translation. Furthermore, the amino acid sequence undergoes a process called folding, which results in a protein with a three-dimensional structure. Once the amino acid sequence is determined, the three-dimensional structure of the protein is basically determined spontaneously, and it is believed that the function of the protein is expressed based on this structure. This is called Anfinsen's dogma (Anfinsen et al. 1961; Anfinsen 1973). It is also known that there is a process of reverse transcription from the RNA sequence back to the DNA sequence, which is also shown in Fig. 1.2.

There are 20 amino acids that make up a protein. Therefore, it is necessary to know the combination of the genetic code of DNA corresponding to each amino acid. This combination of DNA and amino acids is shown in the codon table. As shown in Fig. 1.3, three letters of DNA correspond to one letter of amino acid; the set of three letters of DNA is called a codon, and what can be understood from this table is that multiple codons correspond to one amino acid. In addition to the 20 amino acids, there is a stop codon that terminates the sequence. The redundancy of the codons varies from amino acid to amino acid, so their frequency of occurrence is different.

Now, with this knowledge in mind, the biologist answered the physicist's question as follows.

> It is already known that the blueprint of life is in the DNA sequence. What makes living organisms inherently different from matter is that they have a blueprint. Evolution is believed to occur by introducing many mutations into the genome sequence (the entire DNA sequence). Therefore, a complete analysis of the genome sequence would reveal how the blueprint has changed in evolution. It should also be possible to elucidate the causes of many genetic diseases. Of course, it is impossible to fully decode the genome now, but I think we will eventually be able to do it.

	U (T)	C	A	G	
U (T)	Phe	Ser	Tyr	Cys	U
					C
	Leu		Stop	Stop	A
				Trp	G
C	Leu	Pro	His	Arg	U
					C
			Gln		A
					G
A	Ile	Thr	Asn	Ser	U
					C
			Lys	Arg	A
	Met				G
G	Val	Ala	Asp	Gly	U
					C
			Glu		A
					G

Fig. 1.3 Codon table of genetic information. From Mitaku (2015)

1.2 The Impact of the Human Genome Project

Approximately 15 years after the dialog between biologists and physicists, the paper that would lead to the Human Genome Project was published (Wada 1987). The paper argued that the human genome could be sequenced on a realistic budget and in a realistic time frame using robots that could sequence DNA at high speed. The human genome contains more than 3 billion base pairs. With the level of technology available at the time, the genome sequence could not be analyzed without the daily tedious work of many talented researchers. There was an argument at the time that such a waste of talent should be avoided. In contrast, the main idea of this paper was that genome analysis could be realized at a realistic cost and time using robots. In addition, in a primitive form, DNA sequencing machines were actually being developed for full automation (Jordan 1993). Thus, this argument was quite persuasive and seemed to have greatly facilitated the launch of the international Human Genome Project. In 2003, the completion of the human genome analysis was declared.

Since then, what was discussed in that paper has been realized one after another. For example, information processing technology was considered to be very important for genome analysis, and many information scientists actually entered the field of genome science. In addition, centers for genome analysis and related research institutes were established in various countries. Many robots have also been developed and put into use. Genome analysis technologies have been used extensively, and many biological genomes and metagenomes have now been obtained.

Genome sequencing has had a major impact on all areas of research related to biology. For example, in medicine, it is essential to understand the causes of hereditary diseases. Human genome analysis provides information on all genes and their polymorphisms, and the benefits to medicine are immeasurable. In agriculture, there have been major impacts, such as the modification of genes related to various phenotypes of agricultural products. In pharmacy, new drugs were developed for drug targets. Furthermore, in molecular-level research such as biochemistry, molecular biology, and biophysics, structural biology has developed along with genome analysis, leading to more sophisticated research.

In genetics and evolutionary biology, genome sequences of different species can now be compared, and evolutionary paths can be studied in detail. Information science, which has nothing to do with biology, has become a very important part of biological research. Genomes are the very information of living organisms, and informatics has become indispensable for genome sequencing. Additionally, since living organisms, although complex, are made of molecules, disciplines such as physics and chemistry have also come to consider biology as one of the frontiers of research.

Furthermore, genome analysis is also related to the humanities. Since the genome sequence provides information on all genes, it is becoming easier to diagnose disease-causing genetic mutations. Additionally, in the case of familial genetic diseases, one diagnostic information can affect many family members. In addition, there are many cases where diagnosis can be made, but treatment is difficult. Genome analysis has raised important "bioethical" issues related to medicine (Little 2002). In this sense, the analysis of the human genome is becoming the interface between the humanities and the material sciences.

1.3 Two Faces of Biology and Two Research Approaches

As mentioned in the previous section, genomes are related to a very wide range of fields, each of which deals with a very diverse set of phenomena. However, we will try to consider issues related to genomes as generally as possible, rather than those of each of these disciplines.

First, we would like you to look at Fig. 1.4. The top half of the figure shows biological individuals, cells, and biopolymers, all of which are entities existing in a three-dimensional real space, each of which is very diverse. Moreover, they are characterized by a complex hierarchical structure. The top half of Fig. 1.4 shows a prokaryotic cell and a human as a representative of a eukaryotic cell and its hierarchy. This is a very simplified model diagram. In actual multicellular eukaryotes, there is a deep hierarchical structure, such as individual-organ-tissue-cell-organelle--biopolymer. As seen in the myoglobin shown here, even the protein at the lowest level has a very complex structure. The complexity increases as one moves up the hierarchy. Therefore, current biological research generally takes a bottom-up approach, starting from the three-dimensional structure of proteins at the lowest level and building up to cells and individuals.

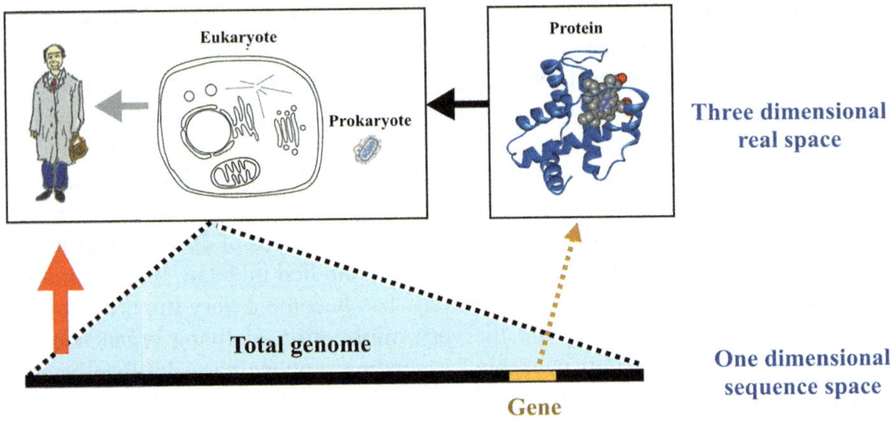

Fig. 1.4 Two ways of looking at organisms: the genome sequence of an organism is a simple one-dimensional array of genes, while an organism in real space is a complex three-dimensional hierarchical structure. Partially from Mitaku (2015)

On the other hand, the lower half of Fig. 1.4 shows the structure of the genome, the blueprint of life. The genome is a one-dimensional sequence of DNA, a collection of all genes. The genes then design the proteins that are the elements of life, and the proteins are combined to form life. In other words, the flow of genetic information is from the genome to the organism, as indicated by the green and red arrows. The genome, a one-dimensional sequence, is a much simpler entity than a three-dimensional organism. In a genome, many genes are simply nested within a one-dimensional sequence. Given this simplicity of the genome, we may be able to solve the biological conundrum by finding simple rules that are not yet known in the one-dimensional arrangement of the genome.

Although the development of genome analysis techniques took a great deal of time and money, it is now much easier to analyze entire genomes. The genomic information of a single species can be stored in the memory of a computer. Since all genes are arranged in a genome sequence, analyzing the genome sequence provides the amino acid sequences of all proteins at the same time. Because of its simplicity, many researchers believe that the genome sequence is merely a catalog of proteins, the parts of an organism. Given the complexity of organisms, it is understandable that they feel compelled to analyze all organisms using a bottom-up approach.

In contrast, there is the idea that the genome sequence is not just a catalog of proteins but that there may be definite rules that are not yet known. In other words, the genome sequence may contain rules that include the harmonization of all proteins, and these should be rules that cannot be inferred from real space. In that case, the rules must be found from the entire genome sequence, and we need to consider the genome sequence in a top-down approach. The physical entity of an organism is made up of four dimensions of space and time, which are converted from a one-dimensional genome sequence. Then, the problem of overall harmony could be hidden in the method of conversion. The history of whole genome analysis in the hands of mankind is young, and the true meaning of the genome as a blueprint may still be poorly understood.

The genome sequence, the blueprint of life, has two faces. On the one hand, it is the face of a collection of genes. The genome sequence catalog proteins must be studied using a bottom-up approach. On the other hand, we believe that the whole-genome sequence has a face: a mechanism that harmonizes the parts of the organism. In the Part II, we summarize the current status of biological research using the bottom-up approach, and in the Part III, we introduce our research using the top-down approach.

Before moving on to Parts II and III, however, we would like to show in the next chapter that there are still major unresolved issues regarding organisms between biology and physics.

References

Anfinsen CB (1973) Principles that govern the folding of protein chains. Science 181:223–230. https://doi.org/10.1126/science.181.4096.223

Anfinsen CB, Haber E, Sela M, White FH Jr (1961) The kinetics of formation of native ribonuclease during oxidation of the reduces polypeptide chain. PNAS 47:1309–1314. https://doi.org/10.1073/pnas.47.9.1309

Avery OT, Colin M et al (1944) Studies on the chemical nature of the substance inducing transformation of pneumococcal types. J Exp Med 79:137–158

Jordan B (1993) Traveling around the human genome. INSERM John Libby Eurotext, Montrouge, France

Little P (2002) Genetic destinies. Oxford University Press

Mitaku S (2015) A modern approach to biological science. Kyoritsu-Shuppan Co., Tokyo (in Japanese)

Sadava D et al (2008) Life, 8th edn. Sinauer Associates, Sunderland, MA

Wada A (1987) Automated high-speed DNA sequencing. Nature 325:325–326

Watson J, Crick F (1953) Molecular structure of nucleic acids: a structure for deoxyribose nucleic acid. Nature 171:737–738

Chapter 2
The Relationship Between Biology and Physics

Keywords Phase diagram of life · Mutation neutrality · Stable state of life · Low entropy state · Convergent evolution

If we look at individual molecules, the three-dimensional structure of DNA and proteins does not violate the laws of physics. However, if we look at the genome sequence from the elementary process of evolution, random mutations are occurring in each nucleotide. Therefore, the genome sequence is always changing toward randomization, which means that the genome sequence is the result of chance. Genome sequences are not considered to have any physical inevitability, and the laws of physics are rarely used when discussing genome sequences. However, in the evolution of organisms as a result of many mutations, it is often observed that they follow sophisticated physical laws. Through random mutations in the genome sequence, there may be some rules that allow organisms to discover the laws of physics. Furthermore, it is certain that organisms have maintained a low-entropy state throughout their evolutionary process of over 3 billion years. In other words, organisms have always defied the law of increasing entropy. In this regard, there should be an explanation of the mechanism that is understandable from the physicist's point of view. Furthermore, mutations involved in evolution are generally considered neutral (Kimura 1983). It is also an interesting question whether the neutrality of mutations can also be explained by physics. In this chapter, we discuss these questions and make them the subject of this entire book.

2.1 Insects that Flourished with the Rules of Fluid Mechanics

Gliders and airplanes were designed to mimic large birds gliding through the sky. Insect flight, on the other hand, is quite different from that of large birds. In recent years, high-speed cameras and computer simulations have begun to reveal how insects fly. Insect wings are small compared to their body size, and if they fly like

S. Mitaku, R. Sawada, *Evolution Seen from the Phase Diagram of Life*, Evolutionary Studies, https://doi.org/10.1007/978-981-97-0060-8_2

large birds, they will fall. For this reason, most insects, such as bees and dragonflies, fly by flapping their wings at high speed.

According to images taken by high-speed cameras, bees flap their wings more than 180 times per second, enabling them to hover. At this time, the insect flaps its wings by twisting the base of the wings. According to hydrodynamic simulations of this flapping, turbulence is generated on the upper side of the wings at a certain moment when the wings are twisted and flapped, generating an upward lift force. By generating this moment with high frequency, the insect maintains lift. This is why insects can hover stationary in the air (Sane 2003; Kolomenskiy et al. 2019). Figure 2.1 very simply shows the location of the turbulence around the wings that occurs when an insect flaps its wings. This method of insect flight has only recently been understood. However, insects hydrodynamically mastered this difficult flight technique hundreds of millions of years ago and have thrived by inheriting it. Their prosperity is evident in the fact that insects account for nearly half of all living species.

Furthermore, insects and birds are very distant in terms of evolution, although small birds such as hummingbirds are said to fly like insects (Warrick and Tobalske 2005). In other words, these organisms have mastered the laws of physics on their own. In biological terms, such evolution is called convergent evolution, meaning that they converge on the same physical laws. Furthermore, while some may think that biology and physics are completely unrelated disciplines, in fact, there are no biological phenomena that are inconsistent with physics. As C. Cockell stated, physics controls life from behind the scenes (Cockell 2017, 2018).

This raises questions about the relationship between genomes and physics. Insects have acquired a way of flight that strictly follows the laws of physics. The number of genes that make this method of flight possible is likely quite large. Are they truly random? Or is there some physical law that governs the harmony of these genes? Or is there some physical law-based pathway for mutations as well? If there is a way to analyze genome sequences using a top-down approach, we should be able to answer such questions.

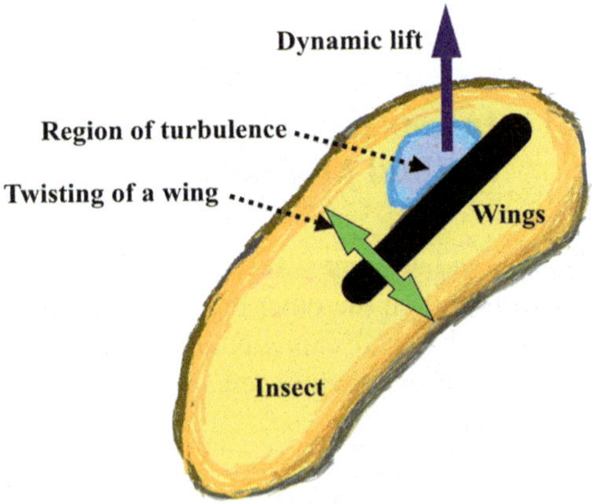

Fig. 2.1 When insects hover, they flap their wings at high speed with a twitch. They use a very sophisticated hydrodynamic mechanism to generate lift, and they have acquired many of the genes necessary for this in the course of evolution

2.2 Order in Organisms and Neutrality of Mutations

In the mid-twentieth century, Schroedinger (1944) wrote his book "What is Life? - A Physical View of the Living Cell." He then raised a major question about the relationship between life and physics. In a closed system, there is the second law of thermodynamics, the so-called law of increasing entropy, which states that order is gradually lost and leads to disorder. In contrast, living organisms consistently exhibit an ordered structure throughout their evolutionary process. Since living organisms are not closed systems, they do not violate the laws of physics to maintain a state of low entropy. However, this mechanism should be physically understandable. Moreover, not only do organisms maintain a low-entropy state, but entropy appears to be even lower in higher organisms. Later, J. Monod pointed out the same problem in his book "Chance and Necessity" (Monod 1971). If we carefully consider the question of how organisms have maintained order, this problem is still not fully understood, even in the current genomic age.

Organisms are highly ordered collections of molecules, including the double helix structure of DNA and the three-dimensional structure of proteins. The DNA sequence of the genome designs the amino acid sequence of all proteins. The A-T and G-C complementarity ensures that the amino acid sequences of proteins are accurately replicated and passed on to the next generation, allowing the organism to maintain a stable and ordered structure. However, it is known that random mutations occur in organisms all the time, disrupting this order. At the molecular level, random mutations occur at a certain rate at each nucleotide in the genome sequence. The question is how this random mutation affects the stability of the organism.

There is another fact about the survival of organisms through mutation. Currently, mutations are generally considered to be neutral. That is, mutations are not fixed because they are beneficial to the survival of the organism, but mutations are fixed despite their neutrality (Kimura 1968). This idea was put forward by M. Kimura in 1968, and a year later, a paper was published by J. L. King and T. H. Jukes that supported it even more strongly (King and Jukes 1969). Although there has been considerable debate since then, the idea of neutral mutation is now widely accepted. Furthermore, the idea that even the birth of a new species, in which many mutations occur in the genome sequence, is dominated by neutral mutations has been proposed by Saitou (2009). This implies that many mutations introduced throughout the genome are also generally neutral. We believe that new analytical methods for the entire genome sequence are needed to solve the problem of maintaining biological stability through the accumulation of neutral mutations.

In physics, the state of a system is generally classified as "stable," "neutral," or "unstable." However, in biology, we are faced with the seemingly contradictory problem of being "neutral" in elementary processes and "stable" as a system. We believe that the problem of low-entropy states in living organisms, mentioned earlier, is because the relationship between the neutrality of random mutations and the stability of living organisms is not clear.

Table 2.1 Similarities between matter and living organisms: macroscopic stable states and microscopic random processes

	Macroscopic states	Elementary process
Matters	The three states of matter	Random motion of molecules
Organisms	The state of life	Random mutations
Common features	Stable	Random

We will discuss this issue in more detail in Part III, but here we would like to briefly explain it. Table 2.1 compares the elementary processes and macroscopic stability of matter and living organisms. Both matter and living organisms are random processes as elementary processes. That is, in matter, it is random molecular motion, and in living organisms, it is random mutation. On the other hand, macroscopic systems are extremely stable under given conditions as the three states of matter (gas, liquid, and solid) in matter. All living organisms maintain a stable state of life. Thus, matter and organisms have similar properties. In Part III, we will explore this similarity further.

2.3 The Problem of Too Many Combinations

To discuss the relationship between genome sequences and physics, we must consider the problem of the number of combinations. This is because the problem of the number of combinations is important in considering the cause of the low-entropy state of organisms. The human genome consists of 3.2 billion base pairs. Even the *E. coli* genome has over four million base pairs. Furthermore, a single gene usually has approximately 1000 base pairs. When translated into the amino acid sequence of a protein, even a small protein consists of approximately 100 amino acid residues. Therefore, let us take the number of combinations of amino acid sequences in 100 residues as an example.

As shown in Fig. 2.2, there are 20 types of amino acids. There are many more chemicals commonly referred to as amino acids. We do not know why these 20 types of amino acids were chosen. However, the 20 types contain amino acids with various physical properties. Figure 2.2 shows a simple classification of the 20 amino acids by physical properties. There are amino acids with polar groups, amino acids with positive or negative charges, hydrophobic amino acids, glycine without side chains, cysteine, which can form covalent bonds in its side chains, and proline, an imino acid. As shown in the codon table in Fig. 1.3, three nucleotides correspond to 20 different amino acids. This allows us to design amino acid sequences with various physical properties depending on the DNA sequence.

Therefore, if we consider the number of combinations of amino acid sequences of 100 residues, it is approximately 10^{130}. In other words, the amino acids in a functioning protein are chosen from a frighteningly large number of combinations. Moreover, the number of combinations in the entire genome sequence is much

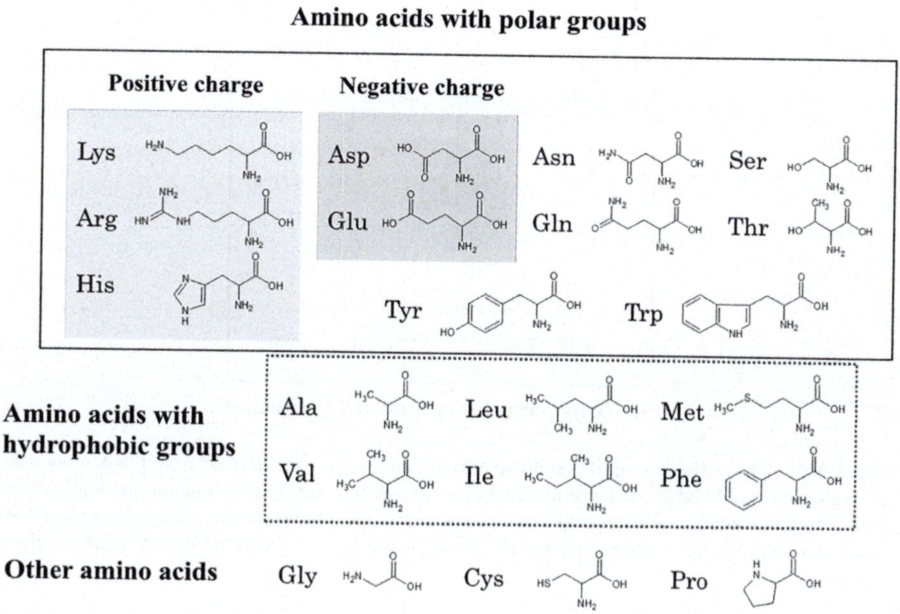

Fig. 2.2 Twenty amino acids used in proteins. From Mitaku (2015)

greater than that for individual proteins. In biology, it would not make sense to consider such a large number of sequences, but in the physics of matter, dealing with a very large number of combinations often provides an understanding of macroscopic properties. In the same way, it would make sense to consider a large number of combinations to consider the macroscopic properties of living organisms.

It is estimated that there are approximately ten million species of organisms worldwide (Trefil 1991). The number of individual organisms (or their genomes) is even larger. However, compared to the number of combinations of DNA of the same length as the genome, it is certainly a very small fraction. Therefore, what are the physical characteristics of the genome sequences that design organisms among DNA sequences of equal length? If it is easy to separate them, we should be able to call it the "phase diagram of life." If we can draw a "phase diagram of life," we should be able to solve the various questions mentioned above.

2.4 Top-Down Approach to Genome Sequences

Figure 2.3 is an illustration of the phase diagram of life. In Part III, we will actually analyze the genome sequence to show that there is a phase diagram of life. At this stage, however, we want to discuss what we can say under the assumption that there is a phase diagram of life. Therefore, suppose that some parameters were extracted from the DNA sequence to describe the phase diagram of life. The parameters must

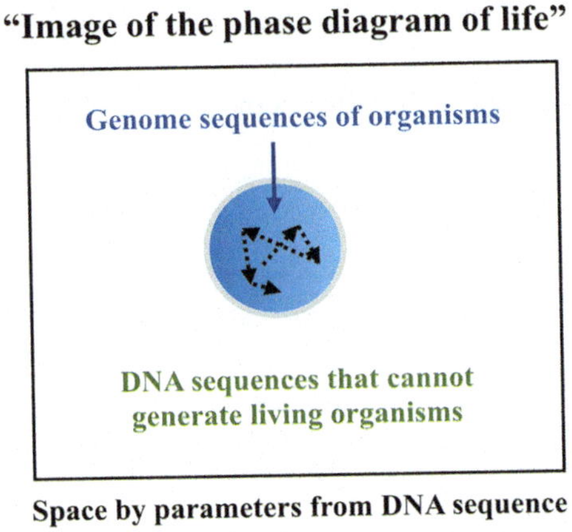

Fig. 2.3 If a phase diagram of life could be drawn using parameters derived from DNA sequences, all organisms would be plotted within that narrow phase diagram. This would automatically eliminate many unwanted sequences and make many mutations neutral. Additionally, organisms walk randomly within the narrow phase diagram, leading to increased diversification. From Mitaku (2015)

have the following characteristics. The number of parameters can be more than one but not too many. When the genome sequences of real organisms are plotted in the space defined by these parameters, all genome sequences will be plotted in a narrow region. In such a case, we can call it a phase diagram of life.

As mentioned in the previous section, the number of genome sequences of all organisms is negligibly small compared to the number of possible combinations of DNA sequences, so if we choose the right parameters, we have a good chance of obtaining a phase diagram of life. If we can draw a phase diagram of life, as shown in Fig. 2.3, the genome sequence will walk randomly without ever leaving the narrow region of the phase diagram. In other words, the genome can produce new genomes while avoiding many useless sequences. The organism will produce new species without ever leaving this region.

As discussed in Sect. 2.1, insects fly with small wings relative to their bodies but are able to perform stable flight maneuvers such as hovering. From a hydrodynamic point of view, this is a very reasonable way to fly, and various genes must be involved in this way of flying in insects. The problem is that it is not easy to arrive at such a combination of genes through a series of completely random mutations. However, the situation is quite different when the genome of an organism is on a random walk in a very narrow region of the phase diagram of life, as shown in Fig. 2.3. If it is a random walk in a very narrow region, it should be easy to find a landing place.

The coexistence of mutation neutrality and low entropy can also be easily solved with a phase diagram of life. If the genome sequence of an actual organism were in a very narrow region on the phase diagram of life, that alone would greatly reduce entropy. The degree of entropy reduction would depend on the smallness of the region and the bias from completely random mutations. In addition, if it is biased away from a completely random state, then from a physics standpoint, there is some conservation law. That conservation law would be related to the maintenance of the state of life. Since life cannot be maintained outside of that narrow region, many useless sequences are automatically eliminated within the phase diagram of life. In other words, mutations are considered neutral in the phase diagram of life.

In Part II, we would like to present the findings from the bottom-up approach, that is, current biology. If the reader is familiar with the contents of Part II, he/she may skip it and proceed to Part III. However, we would like to remind the reader that there are parts of Part II that are to some extent related to Part III.

References

Cockell SC (2017) The law of life. Phys Today 70(3):42–48

Cockell SC (2018) The equation of life: the hidden rules shaping evolution. Atlantic Books

Kimura M (1968) Evolutionary rate at the molecular level. Nature 217:624–626

Kimura M (1983) The neutral theory of molecular evolution. Cambridge Univeristy Press

King JL, Jukes TH (1969) Non-Darwinian evolution: most evolutionary change in proteins may be due to neutral mutations and genetic drift. Science 164:788–798

Kolomenskiy D et al (2019) The dynamics of passive feathering rotation in hovering flight of bumblebees. J Fluids Struct 91:1–18

Mitaku S (2015) A modern approach to biological science. Kyoritsu-Shuppan Co., Tokyo (in Japanese)

Monod J (1971) El hansard et la necessite. Alfred A. Knopf, Inc., Paris

Saitou N (2009) From selectionism to neutralism: paradigm shift of evolutionary studies. NTT Publishing Co., Tokyo (in Japanese)

Sane SP (2003) The aerodynamics of insect flight. J Exp Biol 206:4191

Schroedinger E (1944) What is life? In: The physical aspect of the living cell. Cambridge University Press, Cambridge

Trefil J (1991) 1001 things everyone should know about science. Doubleday, New York

Warrick DR, Tobalske BW, Donald R, Powers DR (2005) Aerodynamics of the hovering hummingbird. Nature 435:1094–1097

Part II
Diversity of Organisms in Real Space

Chapter 3
Three-Dimensional Structures of Proteins Responsible for Biological Functions

Keywords Three dimensional structure · Protein folds · Fluctuation · Molecular recognition · Function of proteins

The smallest structures in the hierarchy of life are biopolymers. Among them, proteins encoded by DNA are fundamental. In this chapter, we will summarize the basic facts about proteins. Proteins are formed by the folding of amino acid sequences into three-dimensional structures. If you look at proteins in the coarse-grained ribbon model, they are surprisingly simple and made up of the folding of a small number of units (secondary structures). There are only approximately 1000 different folding patterns. They fluctuate quite dynamically, and it is thought that the large fluctuating parts form the active sites of molecular recognition.

3.1 Proteins Are Hierarchically Structured

Proteins are biopolymers consisting of amino acids bound together in one dimension (Branden and Tooze 1998). To illustrate how the amino acids are connected, Fig. 3.1 shows the sequence of three amino acids. The left end of the figure is the amino terminus ($NH3^+$), and the right end is the carboxyl terminus (COO^-). The amino acid sequence is synthesized from the amino terminus to the carboxyl terminus. In the diagram in Fig. 3.1, each amino acid has a side chain indicated by R1, R2, and R3. There are 20 types of amino acids that make up proteins, and R1, R2, and R3 refer to one of the side chains. Although only three amino acids are shown here, actual proteins have tens to thousands of amino acids connected to them. In addition, the sequence of amino acids differs from protein to protein. As already shown in Fig. 2.2, side chains have various physical properties.

Amino acids are peptides bonded at the NH-CO moiety. Peptide bonds have a planar structure similar to a double bond, and the dihedral angles are fixed. However, an amino acid has two bonds that can rotate freely, and these dihedral angles are represented by φ and ψ. Thus, when several hundred amino acids are connected, many structures are possible, which behave like thin strings. Due to the excluded

S. Mitaku, R. Sawada, *Evolution Seen from the Phase Diagram of Life*, Evolutionary Studies, https://doi.org/10.1007/978-981-97-0060-8_3

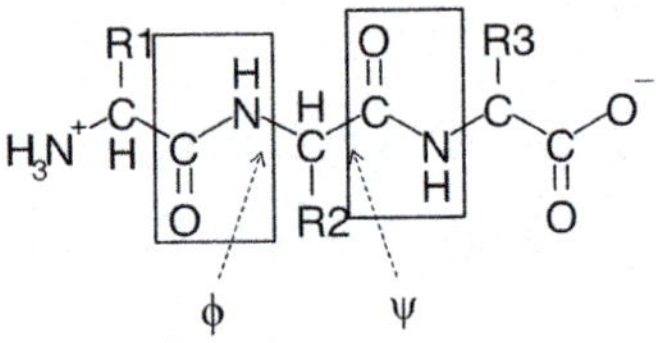

Fig. 3.1 Oligopeptide formed by the bonding of three amino acids. From Mitaku (2015)

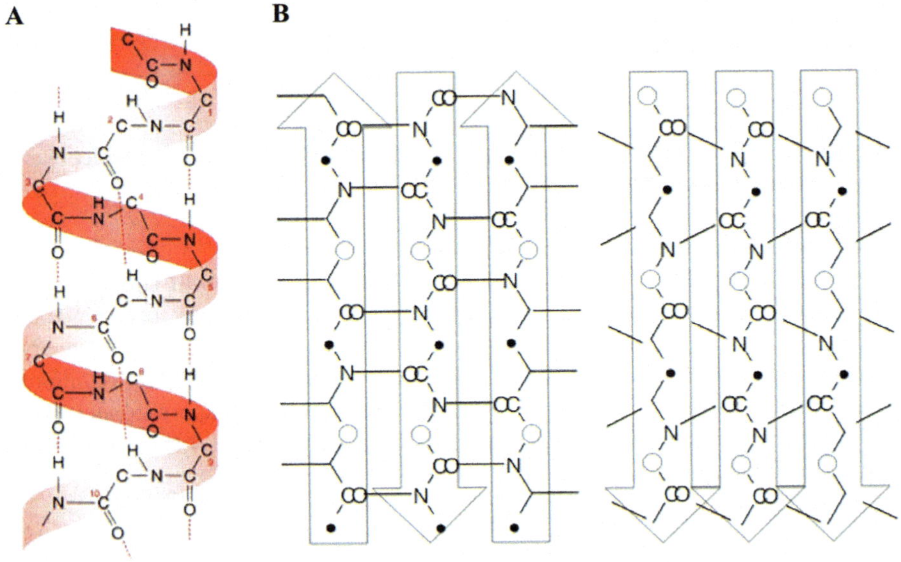

Fig. 3.2 Secondary structure of proteins: α-helix (**a**) and antiparallel as well as parallel β-sheet (**b**). From Mitaku (2015)

volume effect of adjacent side chains, roughly three orientations of the dihedral angles φ and ψ are possible. Thus, for a 100-residue amino acid sequence, approximately 10^{95} possible conformations are possible. However, the number of three-dimensional structures of actual proteins is very small, and they have a fixed shape.

When looking at a protein that has been structurally analyzed, it is clear that its structure is hierarchical. First, the amino acid sequence forms a local structure called a secondary structure. Figure 3.2 shows typical secondary structures, α-helices and β-sheets (parallel and antiparallel types). α-helices are stabilized by hydrogen bonds between the carboxyl group of an amino acid and the amino group of the amino acid four residues ahead. The β-sheet is stabilized by hydrogen bonds between amino acids that are quite far apart. The formation of these secondary structures determines the dihedral angles (φ and ψ) of each amino acid in the area. However, given an amino acid sequence, it is not easy to determine which part will be the α-helix or β-sheet. This is because interactions with the surrounding amino acid sequence and with the solvent are important factors. In the case of membrane proteins, prediction has been possible since the end of the twentieth century

(Hirokawa et al. 1998), but it is only recent that highly accurate prediction has become possible (Jumper et al. 2021; Tunyasuvunakool et al. 2021).

In determining the three-dimensional structure of a protein, it is not only the well-structured secondary structure that is important but also the loose structures called turns and loops at the ends of the secondary structure. These structures are often found on the surface of the protein and in the active site, where there is much fluctuation. Figure 3.3 shows combinations of amino acids commonly found in turn and loop structures (Imai and Mitaku 2005). Proline alone can disrupt the secondary structure, and when two glycines are connected, the secondary structure is almost destroyed. Furthermore, highly accurate secondary structure breakers can be predicted if the surrounding amino acid sequence is taken into account. Clusters of small polar side chains, such as serine and threonine, as well as clusters of amphiphilic amino acids, such as lysine, arginine, and glutamic acid, can also act as secondary structure breakers. By combining these factors, more than 90% of the turn structures could be predicted (Imai and Mitaku 2005).

As such, the three-dimensional structure of a protein is composed of a combination of secondary structures and turns. Each local structure is determined by the sequence features of the amino acid fragment, its interaction with the surrounding amino acid sequence, and its interaction with the environment, such as water and membranes. In other words, the three-dimensional structure of a protein is largely determined by designing the amino acid sequence. This is called the Anfinsen dogma (Anfinsen 1973). In summary, the DNA base sequence determines the amino acid sequence, which in turn determines the three-dimensional structure and

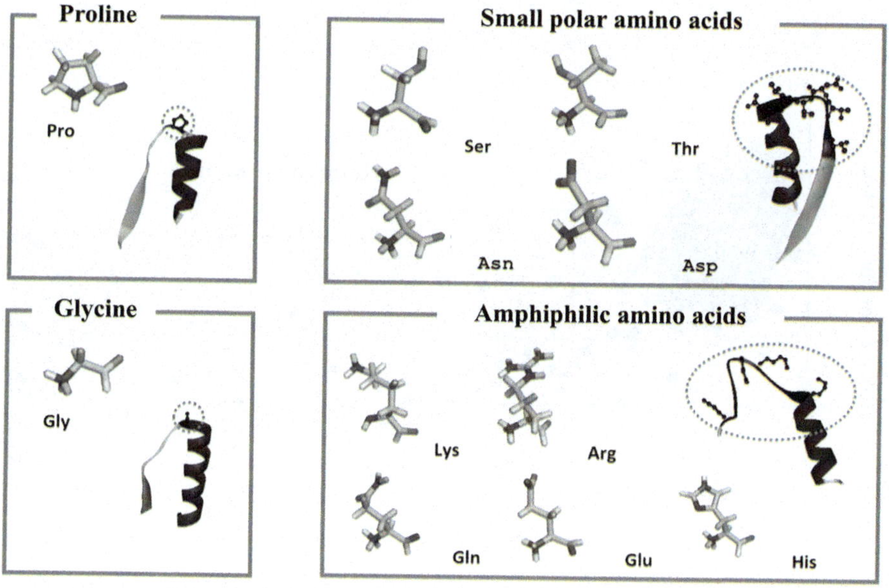

Fig. 3.3 Amino acid sequences commonly found in protein turn structures. From Mitaku (2015)

function of the protein, so the DNA base sequence designs all living organisms at the molecular level.

3.2 Patterns of Protein Conformation

Broadly speaking, protein conformational patterns can be broadly classified into four types according to the combination of secondary structures: alpha, beta, alpha+beta, and alpha/beta types. Alpha and beta types are proteins composed of only their respective secondary structures. The alpha+beta type is a structure in which alpha and beta types are linked together. The alpha/beta type is a structure in which the alpha helix and beta sheet appear alternately.

The patterns are roughly the four described above, but there are many combinations, depending on the number and length of each secondary structure and the order in which they appear in it. How many combinations there are was discussed in the 1990s. The pattern of combinations of secondary structures is called a fold. According to studies on the homology of three-dimensional structures, there are approximately 1000-fold differences (Chothia 1992).

Figure 3.4 shows the relationship between amino acid sequence homology and protein three-dimensional structure homology. Here, pairs of proteins with similar three-dimensional structures were extracted, and the homology (%) of the amino acid sequence and the homology (RMS) of the three-dimensional structure are shown in a scatter plot. This graph is characterized by the presence of many pairs with very low amino acid sequence homology but high conformational homology.

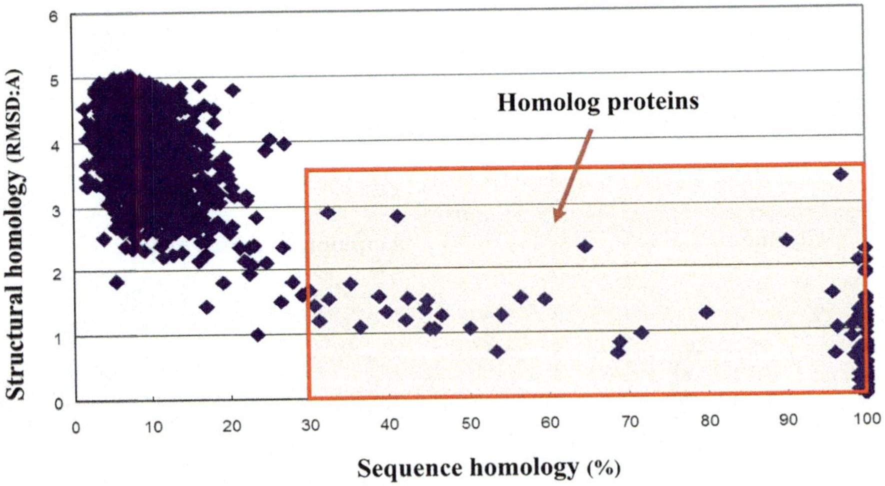

Fig. 3.4 Scatterplot of structural similarity versus sequence similarity for protein pairs with similar 3D structures. From Mitaku (2015)

As mentioned earlier, although the combinations of amino acid sequences are very large, there are only approximately 1000 conformational patterns, called folds (Chothia 1992). In other words, many amino acid sequences converge on the same three-dimensional structure. The question is why protein conformations converge approximately 1000-fold. One of the reasons is that even if the physical properties of amino acid sequence fragments are completely different, they can form exactly the same structure. For example, as shown in Fig. 3.3, there are several types of amino acid fragments that form turns. Proline causes a turn, and two glycines also cause a turn. In addition, a sequence of amino acids that is highly hydrophobic in a membrane protein will form an α-helix. In contrast, in an α-helix exposed to water, such as calmodulin, the sequence of amino acids is highly hydrophilic. Thus, the same secondary structure has completely different amino acid sequences depending on the surrounding environment. Thus, it is physically natural for proteins with similar overall three-dimensional structures to have completely different amino acid sequences.

Thus far, we have considered the three-dimensional structure of proteins, but one of the most important features of proteins is their function. Therefore, we would like to consider the relationship between three-dimensional structure and function. In many cases, proteins with similar three-dimensional structures have similar functions. However, we must be careful because there are cases such as Fig. 3.5. Figure 3.5 shows lipase, which consists almost entirely of α-helices, and trypsin, which consists almost entirely of β-sheets (Mitaku 2015). The former is a lipase that degrades lipids, while the latter is a protein-degrading enzyme. Although their

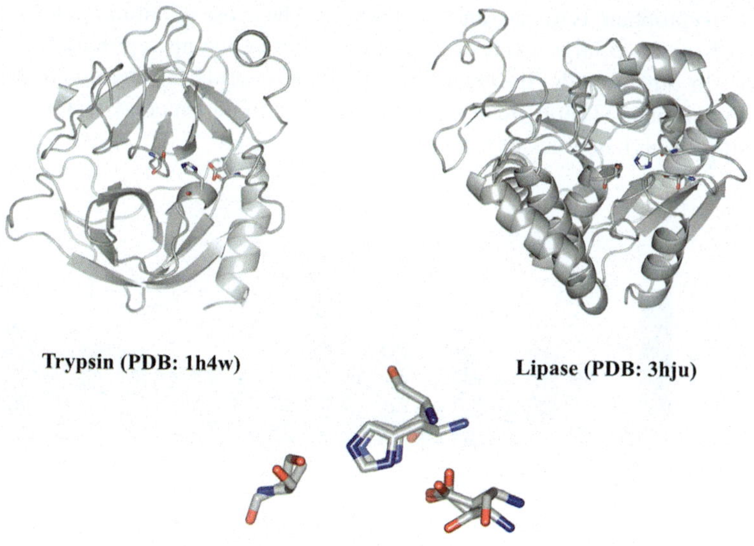

Fig. 3.5 Common active site residues in protein pairs with dissimilar 3D structures: trypsin (PDB:1h4w) and lipase (PDB:3hju). From Mitaku (2015)

overall structures are quite different, their functions are very similar. This is because the three residues in the active site responsible for the degrading function are almost identical in both.

In any case, as seen from Fig. 3.5, the function-determining amino acids are only a small part of the whole. Therefore, the factors that determine the three-dimensional structure of the entire protein can be considered to have little to do with the function-determining amino acids. In other words, when considering folds, amino acid sequences related to function can be considered mostly noise. The amino acid groups that are relevant to function are thought to be due to natural selection. However, the question remains as to how the fold is formed. This issue will be discussed in Part Three.

3.3 Dynamic Changes in the Three-Dimensional Structure of Proteins

Next, let us consider how proteins are able to perform their diverse functions. First, let us look at the distribution of amino acids encoded by a single genome. Figure 3.6 shows the frequency distribution of amino acids used in *E. coli*. The most abundant amino acid is the hydrophobic amino acid leucine, followed by amino acids with small side chains: alanine, glycine, and valine. Smaller side chains reduce the excluded volume and increase the degree of segmental freedom. In other words, the segments become more flexible. On the other hand, less frequent amino acids are cysteine, tryptophan, tyrosine, and methionine. These are all amino acids that form S–S bonds or have large exclusion volumes, thus reducing segmental degrees of freedom. In other words, the occurrence of amino acids that harden segments is generally inhibited. Therefore, it is the nature of genome-encoded proteins to create active sites of high degrees of freedom.

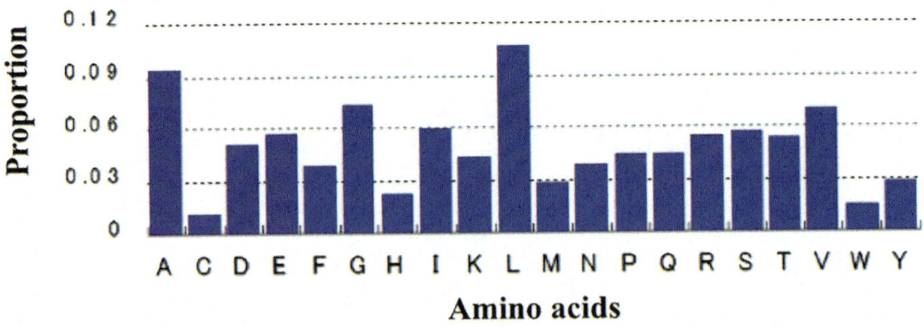

Fig. 3.6 Distribution of amino acid composition in the *E. coli* genome. Hydrophobic amino acids such as leucine, alanine, and valine and segment-softening amino acids such as glycine are abundant. On the other hand, segment-hardening amino acids such as cysteine, tryptophan, and tyrosine are scarce. From Mitaku (2013)

Thus, while proteins are uniquely structured, most protein structures are designed to be flexible and structurally mobile. In general, it is often only small regions, such as the active site, that exhibit large fluctuations, and the entire protein rarely undergoes large conformational changes. Even so, large changes do occur occasionally, and one example is calmodulin.

Figure 3.7 shows two states of calmodulin. Calmodulin is a typical dumbbell-shaped protein, with two spherical subunits connected by a single α-helix. Each spherical subunit has two calcium-binding sites. In the structure on the left, there is no calcium, and it is in the form of an extended dumbbell. However, when calcium binds, it folds as in the structure on the right, involving other proteins. Calmodulin interacts with a variety of proteins, and it is believed that calmodulin thereby regulates the function of those proteins. This conformational change is possible because calmodulin is a flexible protein.

There is research on the softness of calmodulin in terms of the physical interaction between subunits. It has been shown that there is a repulsive force between the two spherical parts of calmodulin, forming a dumbbell structure. When calcium binds, the repulsive force decreases, and the structure collapses significantly (Uchikoga et al. 2005).

Proteins that show large structural changes, such as calmodulin, are not thought to be very common, but proteins move more or less when they function. In degradative enzymes such as lipase and trypsin, the side chains of amino acids at the active site are moving during function. Additionally, in proteins such as transport systems, several amino acid side chains are moving to perform transport. All of these proteins function by recognizing other molecules. Therefore, in the next section, we consider molecular recognition by proteins.

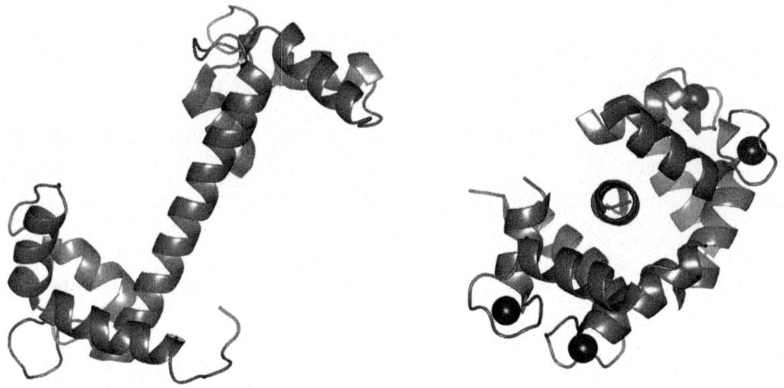

Fig. 3.7 Two types of three-dimensional structures in calmodulin: dumbbell-shaped calmodulin and calmodulin in a form that involves other proteins. From Mitaku (2015)

3.4 Molecular Recognition by Proteins

Calmodulin binds to calcium, which binds to other proteins and turns their functions on and off. To think of protein function more generally, proteins switch various functions on and off by recognizing other molecules. In the case of computers, information is processed by electrical switches. In contrast, an organism can be viewed as an information processing machine, and the basic process is molecular recognition by proteins. This may be a rough analogy, but molecular recognition is the most important biological function of proteins.

Figure 3.8 is a model of a protein showing the relationship between the whole protein and the active site. As discussed in this chapter, there are approximately 1000 different structural patterns (folds). On the basis of these folds, various active sites are formed by the combination of relatively small amino acids, and the corresponding functions are expressed in each of them. Thus, to understand protein diversity, we need to consider two issues. One is the question of how folds are formed. It is thought that there is some physical factor that narrows the fold down to approximately 1000 different types. The other is the question of how to form the active site of function with relatively few amino acids.

Thus, we can think of the problem of protein structure as two separate problems. For the functional site, we consider the fold structure as a given. On the other hand, when we think about the fold, we consider the functional site as a kind of noise. The former approach is better suited for understanding proteins from a functional point of view and has been the subject of much research. However, the general theory in the latter approach has not been given much importance. In Part Three, we will consider the genome as a whole from the latter viewpoint.

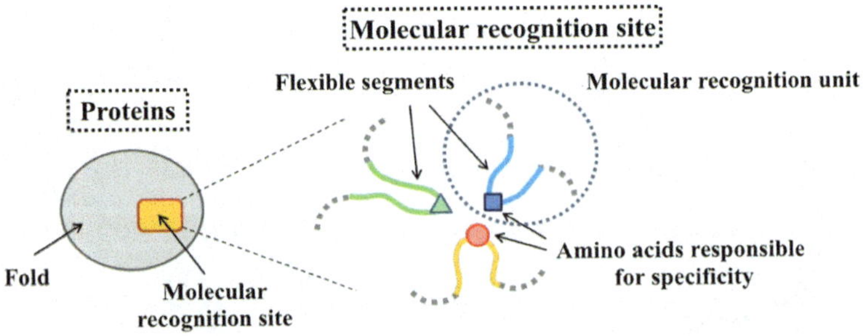

Fig. 3.8 The three-dimensional structure of a protein consists of a molecular recognition site responsible for functional specificity and a supporting fold. The molecular recognition site accounts for a few percent of the total structure, and the fold is composed of all amino acids

References

Anfinsen CB (1973) Principles that govern the folding of protein chains. Science 181:223–230. https://doi.org/10.1126/science.181.4096.223

Branden C, Tooze J (1998) Introduction to protein structure, 2nd edn. Garland Science, New York

Chothia C (1992) One thousand families for the molecular biologist. Nature 357:543–544

Hirokawa T, Seah B-C, Mitaku S (1998) SOSUI: Classification and secondary structure prediction system for membrane proteins. Bioinformatics 14:378

Imai K, Mitaku S (2005) Mechanisms of secondary structure breakers in soluble proteins. Biophysics 1:55–65. https://doi.org/10.2142/biophysics.1.55

Jumper J, Evans R, Pritzel A et al (2021) High accurate protein structure prediction with AlphaFold. Nature 596:583–589

Mitaku S (2015) A modern approach to biological science. Kyoritsu-Shuppan Co., Tokyo (in Japanese)

Tunyasuvunakool K, Adler J, Wu Z et al (2021) Highly accurate protein structure prediction for the human proteome. Nature 596:590–596

Uchikoga N, Takahashi S, Ke R-C, Sonoyama M, Mitaku S (2005) Electric charge balance mechanisms of extended soluble protein. Protein Sci 14:74–80

Chapter 4
Molecular Devices that Support Genome Processing

Keywords Molecular devices · Replication · Translation · Mutation · Repair enzymes

A variety of molecular devices are involved in proteins, which shape their structure and function. Among them, the group of molecular devices involved in genome processing is the most important. First, there are processes such as replication, transcription, and translation that directly process the genetic information written in the genome sequence to form proteins. Chaperones and translocons then form the three-dimensional structure of the protein from the synthesized amino acid sequence. In eukaryotes, cells are large, contain a variety of organelles, and have several intracellular processes. The genome is folded very compactly and stored in the nucleus. Folding of the genome itself is also performed by protein assembly. Then, there is an RNA editing process called splicing. The nuclear membrane contains nuclear pores that specifically transport nuclear proteins synthesized in the cytoplasm into the nucleus. The various mutations and their repair processes are important in terms of finalizing the genome sequence.

4.1 Molecular Devices from DNA Replication to Translation into Amino Acid Sequences

The process from the DNA sequence of a gene in the genome to the amino acid sequence and then to the formation of a functional protein is modeled for a eukaryotic cell, as shown in Fig. 4.1. In eukaryotes, the genome is contained within the nucleus, the largest organelle. The genome is highly folded, forming chromosomes; the double helix of DNA is exposed in a partially dissolved form. Part of this is the gene region. In Fig. 4.1, genes A, B, and C are encoded separately, and models of proteins A, B, and C made from each gene are shown. In general, proteins have different three-dimensional structures and functions and maintain life as a whole.

Many processes and molecular devices are omitted from this diagram but are shown to give an understanding of the overall flow. More detailed processes are

S. Mitaku, R. Sawada, *Evolution Seen from the Phase Diagram of Life*, Evolutionary Studies, https://doi.org/10.1007/978-981-97-0060-8_4

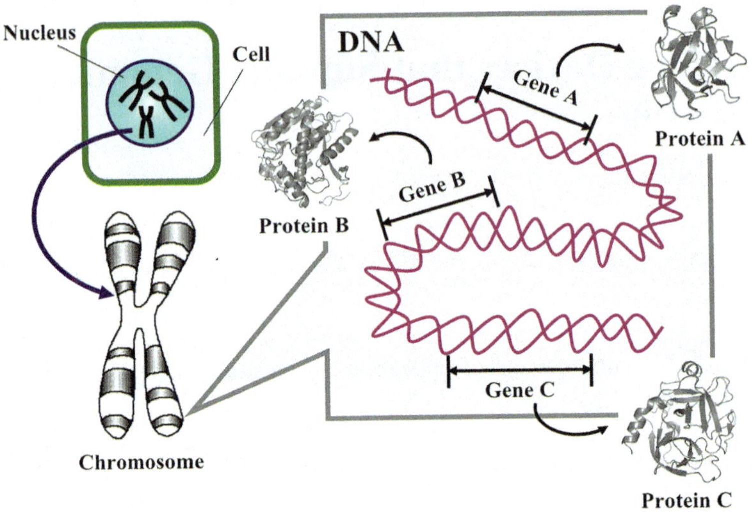

Fig. 4.1 Relationship between cell nucleus, chromosomes, DNA, genes, and proteins in eukaryotes. The genome is located in the DNA folded into the nucleus. Various proteins are formed based on information from various genes in the genome. From Mitaku (2015)

described below. As shown in Fig. 1.1, the two strands of double-helical DNA are complementary to each other and contain the same amount of information. Therefore, if the double helix is unwound and a new complementary DNA is synthesized for each single-stranded DNA, two identical double-helical DNAs will eventually be created: the replication process.

Several molecular devices are used in this process. First, the original double helical DNA must be unwound. Next, DNA polymerase synthesizes the complementary DNA of the unwound single-stranded DNA. Each single-stranded DNA in the double helix is oriented in opposite directions: if the ends of the DNA are 3′ and 5′, one strand is 5′ → 3′ and the other is 3′ → 5′. However, DNA polymerase can only synthesize in the 5′ → 3′ direction. Therefore, one single-stranded DNA can be synthesized in a straightforward manner, while the other single-stranded DNA is synthesized in a complicated process. First, a sequence called a primer is synthesized, followed by the synthesis of multiple DNA fragments, which must later be linked together. This replication requires several molecular devices, including helicases, primases, and DNA ligases.

Next, RNA polymerase is used as the molecular device for transcription from the DNA sequence to the RNA sequence. In this case, however, only the gene region is synthesized. Therefore, a molecular device that recognizes the regulatory region to identify the gene region is needed. This is done by the basic transcription factor complex. DNA replication and RNA transcription are essentially the same in the sense that they utilize base complementarity. The only difference is that RNA uses uracil (U) instead of thymine (T) in DNA. The RNA that encodes the amino acid sequence of a protein is called messenger RNA (mRNA).

Each mRNA is translated into an amino acid sequence by the ribosome. Figure 4.2 shows the role of each molecular device used for this purpose and the flow of synthesis; the molecular devices are complex because translation from an RNA sequence to an amino acid sequence is the conversion of a completely different molecule. In addition to messenger RNA, which encodes the amino acid sequence of a protein, two types of RNA are required for translation. These are ribosomal RNA (rRNA) and transfer RNA (tRNA). The ribosome is an enormous molecular device, a complex of proteins and rRNA. The proteins that make up the ribosome are encoded by genes that are separate from the rRNA genes. The tRNA then binds to each amino acid to form aminoacyl-tRNA.

Then, in forming the amino acid sequence of the protein, the mRNA first binds to the ribosome, and aminoacyl-tRNAs matching the sequence of codons in the mRNA sequentially bind to the already formed amino acid sequence. Finally, the addition of amino acids ends at the terminal codon, completing the amino acid sequence. These translation processes take place in the cytoplasm outside of the nucleus. Thus, in eukaryotic cells, all proteins for molecular devices for replication and transcription must be selectively transported into the nucleus.

4.2 Molecular Devices in the Formation of Three-Dimensional Protein Structures

Synthesis from a DNA sequence to an amino acid sequence alone is not enough for a protein to perform its biological function. It is only when the amino acid sequence is properly folded, forms the active site and moves dynamically that it performs its function. A small protein may be able to form a structure by itself by physical

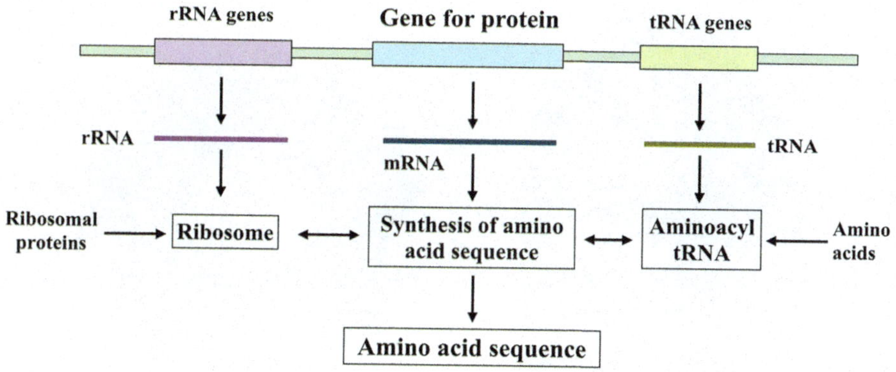

Fig. 4.2 Proteins are synthesized by ribosomes based on information in messenger RNA. Three types of RNA are involved in this process: messenger RNA, which carries information on amino acid sequences; ribosomal RNA, which is part of the ribosome; and transfer RNA, which binds to amino acids

processes. However, cells have molecular devices that help proteins form three-dimensional structures. It is a protein called a chaperone. This is a basket-like molecular device in which many proteins are assembled (Fig. 4.3). Proteins to be folded are incorporated into the chaperone, and the structure is formed. This molecular device, also called a heat shock protein, is able to reform the protein structure even when the body temperature rises and denatures due to infection. This allows pathogens to die, but our bodies are able to survive denaturation.

There is also a molecular apparatus called the proteasome, which works in the opposite way to chaperones, degrading proteins. This system is found in eukaryotes and archaea, and its shape is similar to that of a chaperone. Several ubiquitins bind to the protein to be degraded, and the bound ubiquitin signals the proteasome to degrade the protein. In ubiquitin–proteasome degradation, only certain proteins are degraded. Thus, the impact on the cell is very minimal. Proteins are degraded into small fragments, which are also known to be used for immunity. Chaperones and proteosome are important molecular devices that are functionally quite different but similar in the sense that they are less specific.

Whereas chaperones are involved in the structural formation of soluble proteins, membrane proteins are formed by a different molecular device. That is the system of translocons and related protein groups. Since biological membranes and membrane proteins will be discussed in detail in the next chapter, this section describes the molecular devices involved in the structure formation of membrane proteins (Mitaku 2015).

Synthesis of membrane proteins and soluble proteins begins in the same manner, with the amino acid sequence initially synthesized in the ribosome according to mRNA. In the case of membrane proteins, hydrophobic amino acid segments corresponding to transmembrane helices or signal peptides appear during synthesis. At this point, synthesis is temporarily halted, and SRPs (signal recognition particles)

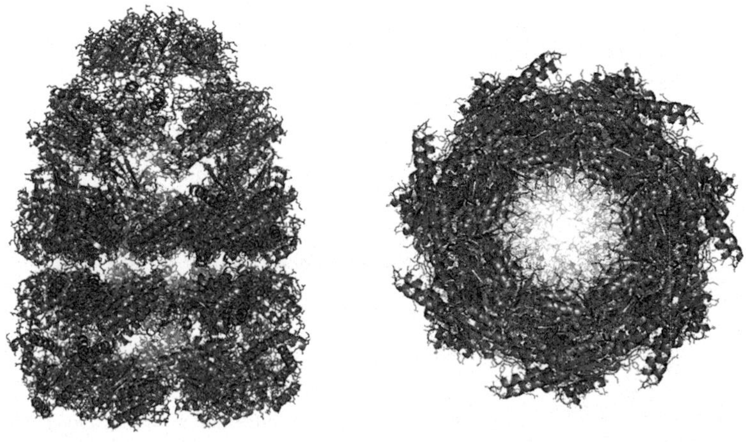

Fig. 4.3 Three-dimensional structure of the chaperone-protein complex. Chaperones have a cage-like structure and act as catalysts for protein conformation. From Mitaku (2015)

bind to the hydrophobic segment. The subsequent process is shown in Fig. 4.4. When the ribosome-SRP complex binds to the membrane translocon, the SRP leaves, and synthesis resumes on the ribosome. Thereafter, the synthesized amino acid sequence is inserted into the membrane. After synthesis, the signal peptide, if present, is detached. Furthermore, the entire amino acid sequence leaves the translocon and becomes a membrane protein.

In the case of membrane proteins, the amino terminus can be either inside or outside the cell, and there may or may not be a signal peptide. Either way, the initial hydrophobic segment will bind to the SRP, which is important for the formation of the protein's three-dimensional structure. Additionally, the driving force to incorporate the amino acid sequence into the membrane seems to utilize the force of synthesis by the ribosome.

4.3 Molecular Devices of the Hierarchical structure of the Genome Itself and the Motor System in the Cell

Eukaryotic genomes are stored in the nucleus and folded tightly. In the case of the human genome, it would be nearly 2 meters long if it were all extended. It is folded into a nucleus that is only a few micrometers long. Moreover, it must be able to unfold as needed, read the genetic information written there, and transcribe it into RNA. In other words, it must be ordered and compacted to almost 1/1000th of its original length, and it must also be able to partially unfold as needed.

In eukaryotic genomes, this is accomplished by hierarchical folding. Figure 4.5 shows the hierarchical structure of the genome. First, a DNA double helix 2 nm thick incorporates a spherical complex of eight histones to form a structure called a nucleosome. The size of a nucleosome is 11 nm. Nucleosomes are composed of a

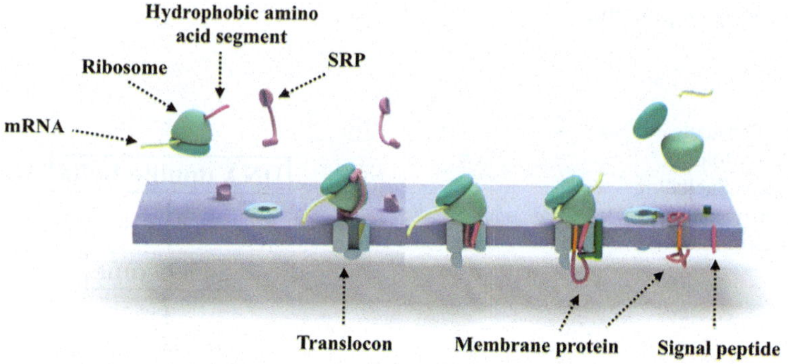

Fig. 4.4 SRPs, translocons, and signal peptidases are involved in the 3D structure formation of membrane proteins. First, SRPs capture the first hydrophobic segments emerging from the ribosome, translocons embed them in the membrane, and signal peptidases cleave signal peptides. From Mitaku (2015)

large number of nucleosomes strung together like a string of beads. Then, the nucleosomes, like a string of beads, are rolled up compactly to form a fiber with a width of 30 nm. This is then folded to form a chromatin structure approximately 300 nm wide. This chromatin is then coiled to form chromosomes. The width of a chromosome is approximately 1400 nm.

This hierarchical structure is formed by almost as much protein as DNA. Those proteins are originally synthesized in the cytoplasm and transported through the nuclear membrane. At this time, proteins that have already formed three-dimensional structures are specifically transported through the nuclear membrane. For this purpose, the nuclear membrane pore has a rather large pore, and a carrier-like protein called an importin is prepared to specifically pass nuclear proteins through this pore. Transported proteins have a positively charged signal sequence called the nuclear localization signal (NLS) (Lange et al. 2007). Importin then recognizes the NLS and specifically transports the nucleoprotein into the nucleus (Fig. 4.6).

The nuclear pore and importin system has an interesting specificity for transporting all the various proteins in the nucleus. The specificity is strict in the sense that it works within the nucleus but very loose in the sense that it allows a variety of proteins to pass through. There are many such molecular recognizers in the genome processing system.

Another transport system involved in genome processing is motor proteins, which move molecules in an organized manner during cell division. Motor proteins include the actin-myosin system and the microtubule-dynein or kinesin system. Actin and microtubules correspond to rails, while myosin, dynein, and kinesin correspond to trains running on rails. Actin filaments and microtubules are long structures composed of polymerized globular proteins. Myosin can move on actin filaments alone, but they polymerize to form thick filaments that form muscles. Dynein and kinesin, on the other hand, can carry and transport cargo alone.

The spindle bodies that emerge during cell division are bundles of microtubules, which move chromosomes. In addition, actin-myosin contracts the middle to split into two cells at the end of division. These motor proteins are essential for cell division in living organisms.

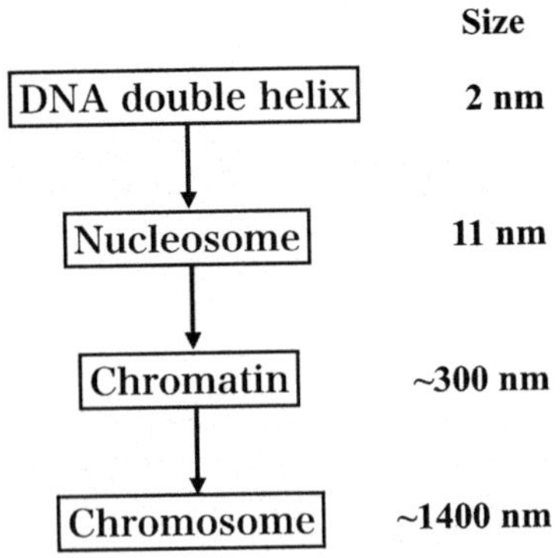

Fig. 4.5 The nuclear DNA double helix folds at several steps, including nucleosomes, chromatin, and chromosomes

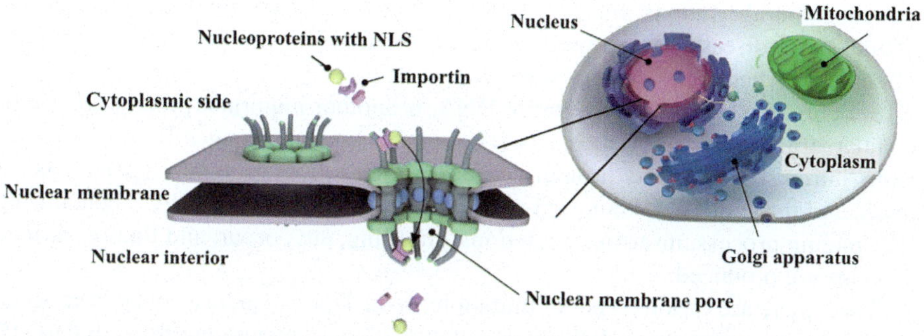

Fig. 4.6 Nucleoproteins synthesized in the cytoplasm are transported specifically into the nucleus through the nuclear membrane pores. In this process, transport proteins called importins bind to and transport nucleoproteins. From Mitaku (2015)

Table 4.1 Different types of mutations

Single-base substitution	A single nucleotide substitution in a nucleotide sequence, which is caused by a duplication error, etc.
Indel	Insertion and deletion of relatively short sequences
Copy number variation	Copy-paste or loss of specific sequences, which is a prominent mutation in higher organisms
Chromosome anomalies	Changes in the length or number of chromosomes

4.4 Mutations to the Genome Sequence and Their Repair Enzymes

We have discussed genomic information as something that should be conserved. However, although the probability is low, various mutations constantly enter the genome sequence. Furthermore, since there are systems that repair mutations, the genome sequence is maintained as a balance between mutation and repair. Therefore, the genome sequences we analyzed are basically the result of that balance. Table 4.1 shows the types of mutations.

The mutations found in all organisms are primarily single-nucleotide substitutions, insertions, and deletions (Sadava et al. 2008). They are caused by errors in replication during cell division or chemical changes in DNA. There are several repair enzymes for these mutations, and in reality, the probability of mutation is very suppressed (Modrich and Lahue 1996; Sancar et al. 2004). However, the repair enzymes are not perfect, and some of the mutations that occur remain. It is also believed that the birth of new species is due to the accumulation of mutations. This process occurs in all organisms, both prokaryotes and eukaryotes.

Indels are a type of mutation in which a relatively short sequence is inserted or deleted. Copy number mutations are more prominent in higher organisms and are mutations in which long sequences are copied and pasted or deleted. For chromosome aberrations, the length or number of chromosomes may be altered. While some diseases are caused by these abnormalities, chromosomal changes often occur when a new species is born.

Splicing, the systematic editing of RNA, is another important process. In multicellular organisms, transcribed RNA is not the final messenger RNA. It has an exon–intron structure, and through a process called splicing, the intron portion is cut away, and the exons are connected to form a mature messenger RNA. Then, during the splicing process, insertions, deletions, shuffling, etc., occur, and finally, various proteins are produced.

Thus, there are various types of mutations in the DNA sequence, and it is believed that evolution occurs through the accumulation of these mutations. Thus, in Parts III and IV, we will discuss in detail the effects of the accumulation of random mutations in the genome sequence and evolution.

References

Lange A et al (2007) Classical nuclear localization signals: definition, function, and interaction with importin a. J Biol Chem 282:5101–5105

Mitaku S (2015) A modern approach to biological science. Kyoritsu-Shuppan Co., Tokyo (in Japanese)

Modrich P, Lahue R (1996) Mismatch repair in replication fidelity, genetic recombination, and cancer biology. Annu Rev Biochem 65:101–133

Sadava D et al (2008) Life, 8th edn. Sinauer Associates, Sunderland, MA

Sancar A et al (2004) Molecular mechanisms of mammalian DNA repair and the DNA damage checkpoints. Annu Rev Biochem 73:39–85

Chapter 5
Biological Membranes and Membrane Proteins

Keywords Membrane proteins · Hydrophobic interaction · Fluidity of membrane · Desaturation experiments · Function of membrane proteins

We focus specifically on biological membranes and membrane proteins here because the physical interactions involved in maintaining their structure are relatively simple. Biological membranes consist of a lipid bilayer in which membrane proteins are embedded. In other words, only amino acid sequences that are physically compatible with the environment of the lipid bilayer can become membrane proteins. Thus, if we first understand the interactions that form the lipid bilayer, we can learn why membrane proteins are embedded in the membrane. Then, to understand the conformational formation of membrane proteins, we need to know about transmembrane helix-helix interactions. An understanding of membrane proteins must lead to an understanding of proteins as a whole. Since many membrane proteins are functionally very important, we decided to focus specifically on membrane proteins.

5.1 Structure of Biological Membranes

All living organisms are composed of cells and have a cell membrane that separates the inside and outside. In addition, eukaryotic cells contain various organelles, such as the nucleus, endoplasmic reticulum, mitochondria, Golgi apparatus, and lysosomes, each of which has its own biological membrane. These biomembranes have essentially the same structure and are composed of a lipid bilayer and membrane proteins, as shown in Fig. 5.1 (Ohnishi 1980). In the lipids that make up the lipid bilayer, a polar group with high affinity for water and two hydrocarbon chains with low affinity for water are covalently linked.

The polar groups of lipid molecules are aligned toward the aqueous phase, and the hydrocarbon chains aggregate to avoid contact with the aqueous phase. Thus, lipids form a thin bilayer structure, as shown in Fig. 5.1. The thickness of the membrane is determined by the length of the hydrocarbon chains and is approximately

S. Mitaku, R. Sawada, *Evolution Seen from the Phase Diagram of Life*, Evolutionary Studies, https://doi.org/10.1007/978-981-97-0060-8_5

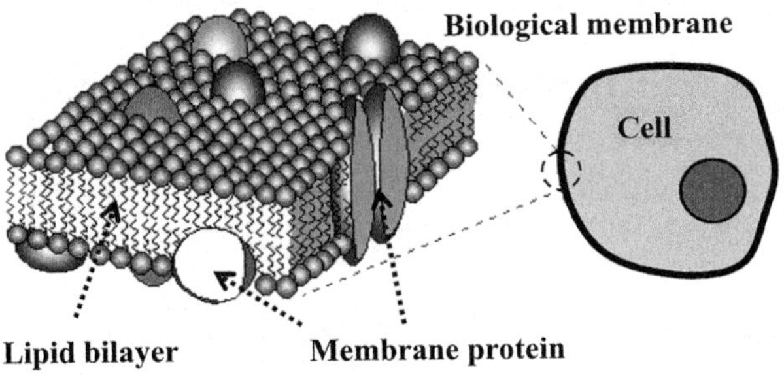

Fig. 5.1 Model diagram of a cell membrane. In general, biological membranes are composed of membrane proteins embedded in a lipid bilayer. From Mitaku (2015)

5–10 nm. On the other hand, the area of the membrane can be essentially infinitely large. In eukaryotes, the size can be 10 μm or more. They can also form very elongated structures, such as the axon of a neuron. Lipid bilayers contain a layer of hydrocarbon chains that repel water, making them less permeable to polar molecules such as water. This property of lipid bilayers and biomembranes allows cells and various organelles to separate and retain polar molecules in and out.

5.2 Hydrophobic Interactions That Form Membrane Structures

The structures of the various molecules and molecular complexes that make up living organisms are formed by physical interactions. Thus far, we have discussed physical interactions such as the formation of the double helix of DNA through hydrogen bonds between bases and the secondary structure of proteins through hydrogen bonds between main chains. In contrast, lipid bilayers are formed by hydrophobic interactions. This interaction plays a very important role in the formation of various structures in biological systems.

In general, when hydrophobic molecules or groups coexist with water, an apparent attraction called a hydrophobic interaction occurs. Water molecules are electrically neutral as a whole but are polarized because they are curved. That is, the side closer to the hydrogen atom has a positive partial charge, and the side closer to the oxygen atom has a negative partial charge. Because of this polarization, water has a high affinity for ions and polar molecules. On the other hand, hydrophobic molecules and groups such as hydrocarbon chains have low affinity for water and exhibit phase separation from water. This phenomenon is sometimes referred to as hydrophobic interaction because it seems as if the hydrocarbon chains are approaching each other by attraction. However, this phenomenon is not due to an actual attraction between hydrocarbon chains but rather an apparent thermodynamic interaction. This apparent interaction then leads to unique properties such as molecular fluidity within the membrane surface (Tanford 1980).

Figure 5.2 shows a model of the arrangement of water molecules around a hydrocarbon chain. It is believed that a cage structure of water molecules is formed around the hydrocarbon chains. Although the water molecules are in dynamic motion, it is easier to understand the experimental results if we consider that statistically they are arranged as shown in the figure. In other words, it is thought that the water molecules form a cage-like network of hydrogen bonds so that their polarization is not directed toward the hydrocarbon chain. This is in good agreement with the fact that this effect is mainly due to the entropy term. We know from detailed experiments on solubility that the magnitude of the hydrophobic effect is proportional to the surface area of the hydrocarbon chains (Tanford 1980).

All proteins can be classified by their shape: spherical soluble proteins, elongated fibrous proteins, and membrane proteins embedded in lipid bilayers. These shapes are similar to the molecular assembly shapes exhibited by lipid molecules and surfactants. Surfactants exhibit spherical micelles, hexagonal phases with rod-like structures, and lamellar phases with stacked membranes (Fig. 5.3). The ability to form molecular assemblies of various structures is a characteristic of hydrophobic interactions. The actual conformation of proteins is formed by various interactions, including hydrophobic interactions. However, the example of the surfactant-water system, which can take various shapes only by hydrophobic interactions, is suggestive in considering the three-dimensional structure of proteins.

Another feature of hydrophobic interactions is the fluidity of biological membranes (Fig. 5.4). Both lipid molecules and membrane proteins, the main components of biomembranes, are known to diffuse two-dimensionally in the direction of the membrane surface (Ohnishi 1980). The diffusion coefficient of lipid molecules is typically 10^{-8} cm^2/s, allowing them to move freely. Larger membrane proteins can also diffuse two-dimensionally with a diffusion constant of approximately 10^{-10} cm^2/s if they are not bound to other molecules. The hydrocarbon chains of lipid molecules are present inside the bilayer to avoid contact with water and can undergo wobbling motion, with a characteristic time of approximately 1 nm. In addition, a single membrane protein can perform rotational diffusion centered

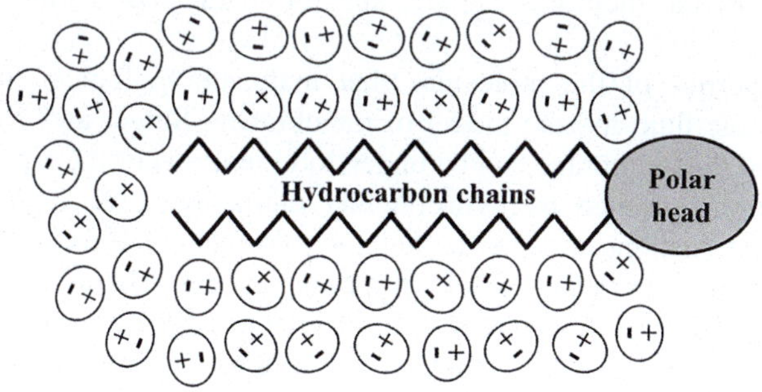

Fig. 5.2 The hydrophobic effect is explained by the arrangement of water molecules. Water molecules around hydrocarbon chains in lipids are thought to be arranged in a cage-like structure

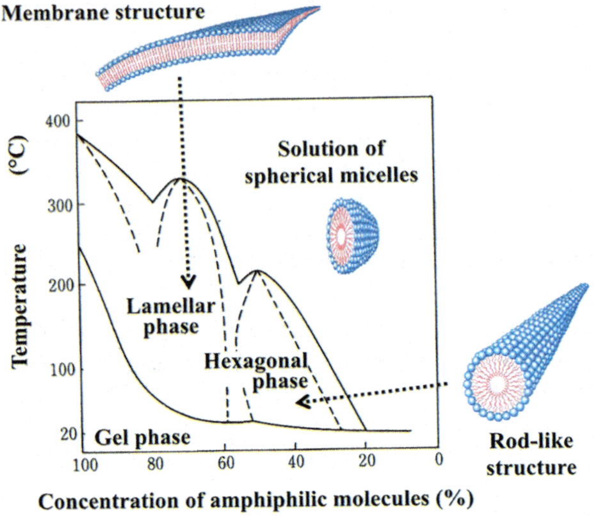

Fig. 5.3 Microphase separation structure of the surfactant-water system. Depending on the surfactant concentration and temperature, structures such as bilayers, rods, and spherical micelles are observed. From Mitaku (2015)

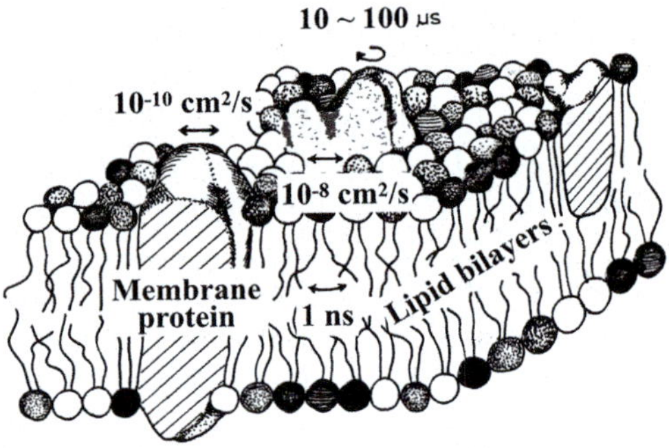

Fig. 5.4 Fluid mosaic model of a biological membrane. Lipids and membrane proteins can diffuse within the membrane surface. From Mitaku (2015)

perpendicular to the membrane plane, and its characteristic time appears to be 10-100 μs.

These properties of the molecules that make up biological membranes are referred to as the fluid mosaic model of membranes (Singer and Nicolson 1972). This fluidity of the membrane can be understood in terms of the fact that the membrane structure is created by hydrophobic interactions. As already mentioned, hydrophobic interactions are apparent interactions that result from the arrangement of water molecules in contact with hydrophobic groups. Therefore, the strength of this interaction does not change unless the state of contact between the hydrophobic group and the water molecules changes. The fluidity of the membrane means that molecular motion that does not change the hydrophobic interaction is free to occur.

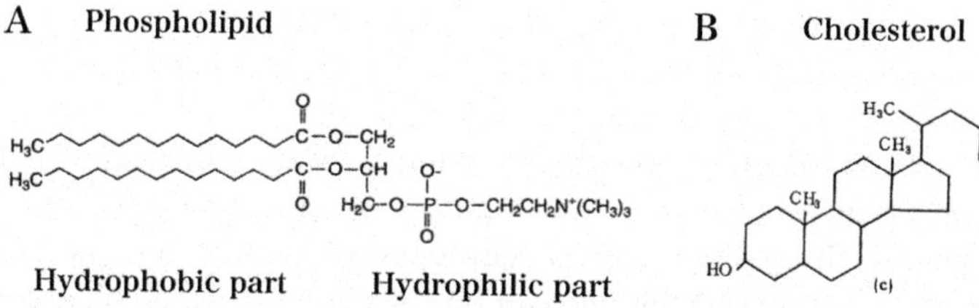

Fig. 5.5 Molecular structures of phospholipid (**a**) and cholesterol (**b**) molecules. From Mitaku (2015)

However, the fluidity and other properties of biological membranes vary considerably depending on the composition of the molecules. Figure 5.5 shows the molecular formulas of typical molecules (phospholipids and cholesterol) that make up the lipid bilayer. Phospholipids have glycerol at the center, two long hydrocarbons at one end, and a hydrophilic group attached at the other end. Phospholipids are those that contain phosphoric acid in the hydrophilic group. Cholesterol is chunky in shape, and lipid membranes containing cholesterol are thought to be stiffer as a whole. If the hydrocarbon chains of phospholipids are uniform in length, the membrane exhibits very sharp structural changes (lipid membrane phase transition) (Mitaku et al. 1983). However, when cholesterol is included, the lipid membrane phase transition disappears. This is thought to be because cholesterol inhibits the degree of freedom of hydrocarbon chains (Sakanishi et al. 1979). This is similar to the fact that when an amino acid sequence contains more bulky side chains, the segmental degrees of freedom are restricted, and the sequence becomes stiffer.

5.3 Interactions That Form the Three-Dimensional Structure of Membrane Proteins

The structure of the transmembrane regions of membrane proteins should be able to be formed by hydrophobic interactions. However, a separate mechanism must be considered to cause the transmembrane helices to assume a specific configuration. Therefore, if we can perform denaturation experiments that disrupt only the tertiary structure while maintaining the transmembrane helix structure, we can estimate the interactions that form the tertiary structure of membrane proteins.

Such denaturation experiments have been performed on several membrane proteins. One such denaturation experiment is the alcohol denaturation of bacteriorhodopsin (Mitaku et al. 1988). Bacteriorhodopsin is a membrane protein present in the plasma membrane of the halophilic bacterium *Halobacterium halobium*. This membrane protein is known to form two-dimensional crystals in the membrane (Oesterhelt and Stoecknius 1971). Bacteriorhodopsin contains a hydrophobic pigment called retinal within the membrane, and when the tertiary structure of bacteriorhodopsin is denatured, the retinal is detached, and the absorption spectrum of light is greatly altered. The changes in secondary and tertiary structure can be examined from the circular dichroism spectrum, and in the alcohol denaturation of bacteriorhodopsin, only the tertiary structure was denatured.

On the other hand, alcohols are easily soluble in water because of their polar groups and can be distributed to some extent in the membrane because of their hydrocarbon chains. Since the partition coefficient of each alcohol is known, the alcohol concentration in the membrane can be calculated if the alcohol concentration in the aqueous phase is known. When alcohol dissolves in the membrane, its hydroxyl groups attack the hydrogen bonds of membrane proteins in the membrane. Therefore, by using alcohols ranging from methanol to butanol and even aldehydes, the concentration of denaturant in the membrane at the time of denaturation can be confirmed. In other words, we can see how many hydrogen bonds are cleaved when the tertiary structure of bacteriorhodopsin is denatured.

Figure 5.6 shows a phase diagram for the denaturation of bacteriorhodopsin with alcohol. The horizontal axis is the concentration of alcohol in the aqueous solution in which bacteriorhodopsin denatures, and the vertical axis is the distribution coefficient of alcohol, shown as a double logarithmic graph. The straight lines in both logarithmic graphs indicate that the concentration of alcohol in the aqueous phase is inversely proportional to the distribution coefficient. In general, the distribution coefficient κ is defined as the ratio of the concentration in the oil phase to the concentration in the aqueous phase, as in Eq. (5.1). Thus, the alcohol concentration C_m in the membrane is constant when bacteriorhodopsin is denatured.

$$\kappa = \frac{C_m}{C_s} \tag{5.1}$$

Quantitatively, experiments showed that the tertiary structure of bacteriorhodopsin is denatured when the concentration of alcohol in the membrane reaches approximately 1 M. In other words, it is estimated that approximately 1 M of polar groups are present in the membrane protein, forming hydrogen bonds and ionic bonds to form the tertiary structure.

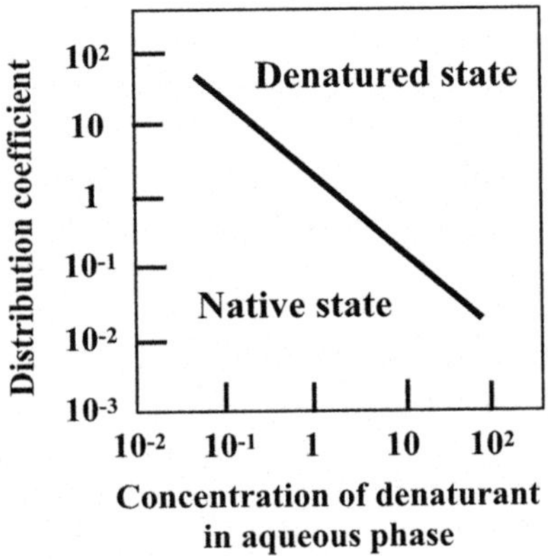

Fig. 5.6 Alcohol denaturation behavior of the membrane protein bacteriorhodopsin. The alcohol concentration at which bacteriorhodopsin denatures is inversely proportional to the distribution coefficient of alcohol

While bacteriorhodopsin is relatively easy to purify and easy to measure spectroscopically, it is generally quite difficult to perform similar denaturation experiments with other membrane proteins. However, we were able to perform similar denaturation experiments on the sodium channels of squid giant nerve axons. Since this membrane protein cannot be studied spectroscopically like bacteriorhodopsin, we investigated its function by membrane potential measurements and performed alcohol denaturation experiments. In physiological studies using squid giant nerve axons, it is common to see how much channel function returns after washing and removing a given chemical. However, in this denaturation experiment, we conducted an experiment to see how much function had been denatured.

The function of sodium channels is to generate action potentials by opening and closing them. By inserting an electrode inside a squid giant nerve axon and perfusing it with a solvent, we can expose the sodium channels to a constant alcohol concentration for a certain time. Then, the percentage of native sodium channels that generate action potentials can be examined (Kukita and Mitaku 1993). The results show that alcohol-induced denaturation of sodium channels is a first-order reaction. Therefore, the concentration of alcohol, which denatures instantaneously as an extrapolation, was almost the same result as in the case of bacteriorhodopsin. This suggests that the tertiary structure of sodium channels in squid giant axons is also formed, as in the case of bacteriorhodopsin, by hydrogen bonding by polar amino acids.

Although off the topic of membrane proteins, some results suggest that in water-soluble globular proteins, stabilization of the globular form is primarily due to hydrophobic interactions, while tertiary structure is due to polar interactions. Globular proteins are said to have a denatured intermediate state called a molten globule. In this state, there is no tertiary structure, the secondary structure is retained, and the whole protein is spherical like a micelle (Fig. 5.7).

To determine whether the molten globule state actually resembles the micelle structure of surfactants, bovine carbonic anhydrase B was denatured with GuHCl (guanidinium chloride), and the distribution of the hydrophobic dye pyrene into the interior was examined (Mitaku et al. 1991). It is known that when bovine carbonic anhydrase B is denatured with GuHCl, it denatures to a molten globule state in a

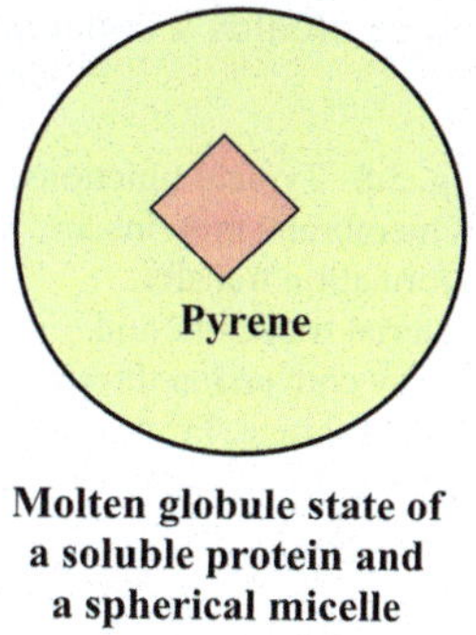

Fig. 5.7 When a hydrophobic fluorescent probe is added to bovine carbonic anhydrase B in a molten globule state, the probe distributes well to the protein. The exact same phenomenon is observed in surfactant micelles

certain GuHCl concentration range. As a comparison, a similar experiment was performed with n-β-octylglucoside to form micelles. The results showed that pyrene exhibited almost the same distribution behavior in protein molten globules and in surfactant micelles (Itoh et al. 1996), suggesting that GuHCl is believed to disrupt polar interactions and that tertiary structures are formed by polar interactions even in globular proteins.

5.4 Functions of Membrane Proteins

Cell membranes have existed since the beginning of life. The plasma membrane has a variety of important functions, most of which are performed by membrane proteins. For example, membrane proteins are responsible for information transfer, material transport, and energy conversion through the membrane (Fig. 5.8). Living organisms (cells) are exposed to enormous amounts of information. That is, they identify information such as nutrients, toxic substances, optical information, sound information, and mechanical contact. Multicellular organisms have developed specialized sensory cells for this purpose. The molecules that sense these stimuli are membrane proteins in the membrane.

Cells take in materials they need and expel materials they no longer need. Membrane proteins responsible for this transport can be broadly classified into active transport, which uses energy, and passive transport, which does not. The former is sometimes called a pump, and the latter is called a channel. The energy used for active transport can come from ATP, the biological energy currency, or from the concentration gradient of ions that are different from the molecule being transported. Passive transport is generally known to be highly specific to the molecule being transported. For example, even sodium and potassium, which have almost identical physical properties other than atomic weight, can be distinguished and transported with high precision. This high specificity is used to generate electrical potentials inside and outside the cell. The electrical signals of neurons are generated by controlling the permeability of ions.

Finally, the generation of ATP is carried out by membrane proteins. First, to synthesize ATP and other high-energy substances, mitochondria metabolize glucose, which creates a concentration gradient of protons (hydrogen ions) across the inner

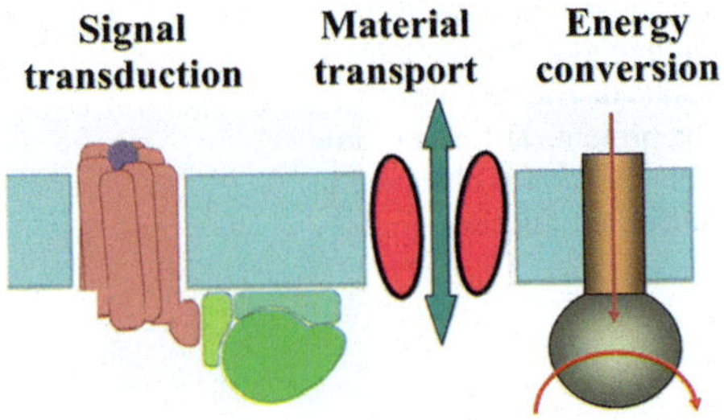

Fig. 5.8 Typical functions of membrane proteins are information transfer, material transport, and energy conversion through the membrane. From Mitaku (2015)

membrane. This proton concentration gradient is the thermodynamic free energy. ATP synthase then fixes this energy into the chemical ATP. This enzyme moves protons along the proton concentration gradient and uses their mechanical energy to synthesize ATP (Itoh et al. 2004).

References

Itoh H et al (1996) Estimation of the hydrophobicity in microenvironments by pyrene fluorescence measurements: n-β-octylgucoside Micelles. J Phys Chem 100:9047–9053

Itoh H et al (2004) Mechanically driven ATP synthesis by F1-ATPase. Nature 427:465–468

Kukita F, Mitaku S (1993) Kinetic analysis of the denaturation process by alcohols of sodium channels in squid giant axon. J Physiol 463:523–543

Mitaku S (2015) A modern approach to biological science. Kyoritsu-Shuppan Co., Tokyo (in Japanese)

Mitaku S, Jippo T, Kataoka R (1983) Thermodynamic properties of the lipid bilayer transition. Biophys J 42:137–144

Mitaku S et al (1988) Denaturation of bacteriorhodopsin by organic solvents. Biophys Chem 30:69–79

Mitaku S et al (1991) Hydrophobic core of molten-globule state of bovine carbonic anhydrase B. Biophys Chem 40:217–222

Oesterhelt D, Stoeckenius W (1971) Rhodopsin-like protein from the purple membrane of Halobacterium halobium. Nat New Biol 233:149–152

Ohnishi S (1980) Dynamic structure of biological membrane. The University of Tokyo Press (UTP), Tokyo (in Japanese)

Sakanishi A, Mitaku S, Ikegami A (1979) Stabilizing effect of cholesterol on phosphatidylcholine vesicles observed by ultrasonic velocity measurement. Biochemistry 18:2636–2642

Singer SJ, Nicolson GL (1972) The fluid mosaic model of the structure of cell membranes. Science (New York, NY) 175:720–731

Tanford C (1980) The hydrophobic effect: formation of micelles and biological membranes, 2nd edn. John Wiley & Sons, Somerset NJ

Chapter 6
Signal Transduction and Enzymatic Metabolic Reactions in Living Organisms

Keywords Signal transduction · Transmembrane proteins · Nerve signals · Enzymes · Allosteric effect

Protein structures are flexible and can change conformation dynamically, allowing them to perform a wide variety of functions. This allows for various phenomena at the molecular level, such as association-dissociation between subunits or allosteric effects due to small molecule binding. In addition to the diversity of protein structures and functions, many proteins form networks and perform more sophisticated functions. Here, we consider the networks of signal transduction that process biological information and the metabolic networks of enzymes.

6.1 How Does Signal Transduction Work?

Organisms receive information from the external environment and respond appropriately to it. External environmental information includes a variety of stimuli, such as light, sound, surrounding mechanical distortions, and a wide variety of chemicals. Multicellular organisms need to communicate with other cells within the same organism. One cell may signal to many other cells, another may signal to only one specific cell. The way signals are sent and received varies from cell to cell (Sadava et al. 2008; Mitaku 2015).

As Fig. 6.1 shows, there are many types of stimuli from the external environment. Many organisms respond to light. These organisms have proteins that receive light. The most common protein that responds to light is rhodopsin, which is a seven-transmembrane helix type. When the retinal in the middle of rhodopsin absorbs light, it isomerizes. The conformational change in the pigment causes a conformational change in the protein, which triggers intracellular signaling.

There are types of receptor proteins that undergo conformational changes upon the binding of chemicals. Most olfactory and taste receptors appear to be seven-helix transmembrane proteins similar to rhodopsin. Different organisms depend on different stimuli from the external environment, and *C. elegans* is known to have a

S. Mitaku, R. Sawada, *Evolution Seen from the Phase Diagram of Life*, Evolutionary Studies, https://doi.org/10.1007/978-981-97-0060-8_6

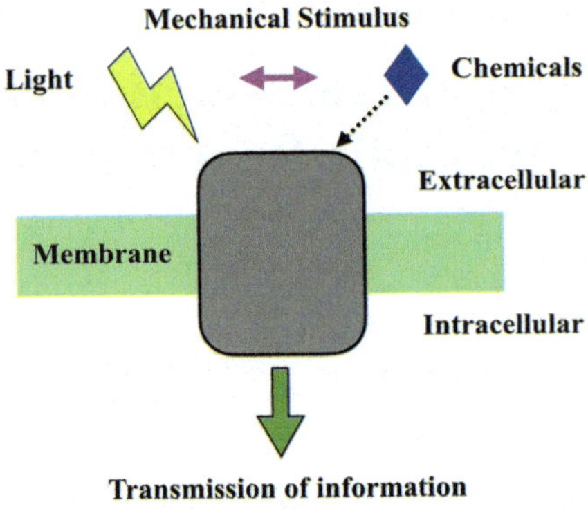

Fig. 6.1 In information transfer at the plasma membrane, light, mechanical, or chemical stimuli cause receptors to change their three-dimensional structure and transmit information to second messengers in the cell

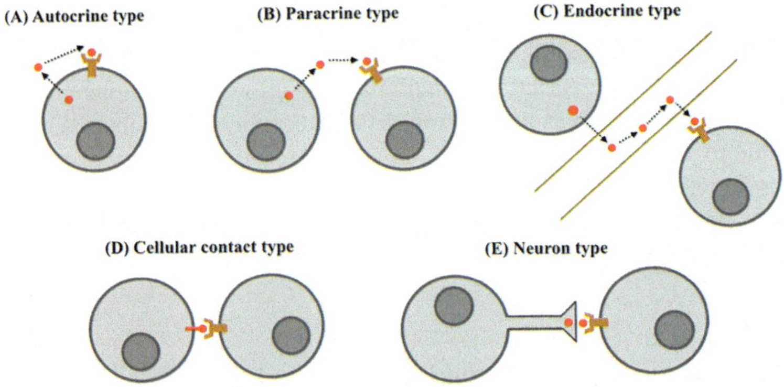

Fig. 6.2 There are various cases of the relationship between transmitting and receiving cells in signal transduction. There are autocrine (**a**), paracrine (**b**), endocrine (**c**), cell junctional (**d**) and neuron (**e**) types. From Mitaku (2015)

wide variety of receptors for chemicals to respond to different chemical compounds. Humans, however, rely heavily on visual information.

Some cells respond to mechanical stimuli. Auditory cells convert sound-induced vibrations into electrical signals on cell membranes. Endothelial cells in blood vessels also deform their shape in response to stress caused by blood flow. Some bacteria seem to sense the geomagnetic field and approach or dive near the surface of the water. They have small magnets in their cells, the movement of which alters the movement of the cells.

In multicellular organisms, intercellular communication is established when one cell releases a chemical substance and various cells receive it. There are various cases depending on the relationship between the transmitting and receiving cells (Fig. 6.2). If the cell that secreted the chemical itself accepts it, it simply amplifies the information; this is called the autocrine type. In some cases, the secreted chemical can only transmit information to the immediate surroundings by diffusion. For example, in the process of wound healing, it is sufficient that only the cells in the immediate vicinity of the wound are altered; this is called the paracrine type.

Since the circulatory system extends throughout the body, it can also transmit information throughout the body. When insulin is secreted from the islets of Langerhans in the pancreas, cells throughout the body respond, and blood glucose levels drop, which is endocrine-type information transmission. Growth factors use a similar mechanism to make the body grow.

On the other hand, cells other than blood cells are attached to each other and recognize each other. As a result, unrestricted cell proliferation is inhibited, and each cell is able to transmit information directly to neighboring cells. In addition, cells themselves may extend their long arms and transmit information only to distant target cells, one example of which is a neuron. In this case, the extended portion is called an axon, and a pouch-like structure called a synapse is formed at its tip. Synapses contain vesicles called synaptic vesicles that contain neurotransmitters and transmit signals to target cells by fusing with the cell membrane. This forms a network with distant cells.

6.2 Information Reception from the External Environment by Seven-Transmembrane Proteins

Many receptors are seven-transmembrane proteins. This type of receptor is found not only in eukaryotes but also in prokaryotic cells. Figure 6.3a shows the three-dimensional structure of rhodopsin, a typical seven-transmembrane protein. The bundle of seven transmembrane helices is almost perpendicular to the lipid membrane. Lysine, located almost at the center of the seventh helix from the amino terminus, is bound to the pigment retinal. In general, a highly polar amino acid such as lysine near the center of the helix would be unstable in a hydrophobic membrane. In

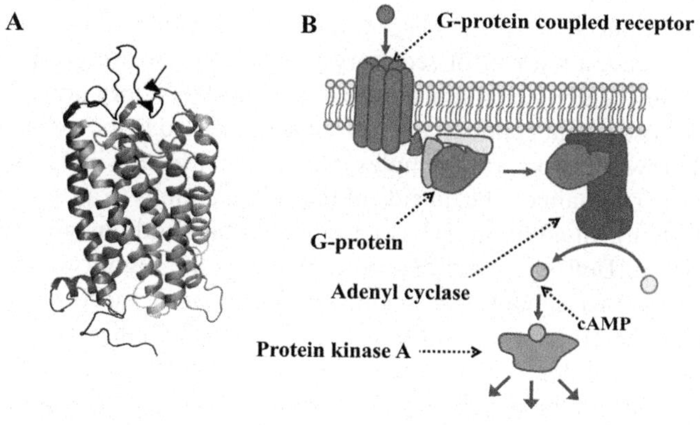

Fig. 6.3 Receptors of seven transmembrane proteins. (**a**) Three-dimensional structure of the visual photoreceptor rhodopsin. (**b**) When a receptor is stimulated, a series of signal transduction reactions occur within the cell. For example, receptor → G protein → adenylate cyclase → protein kinase. From Mitaku (2015)

the case of rhodopsin, however, the first six transmembrane helices seem to contribute to the stability of the last helix in the membrane.

The presence of clusters of hydrophobic amino acids contributes significantly to the stable retention of these transmembrane helices in the membrane. However, it has been found that this alone is not sufficient to stabilize the helix ends at the membrane surface. It is necessary for clusters of polar amino acids to be present near the membrane surface. This can be understood by computational experiments in which membrane proteins are pulled out from their amino or carboxyl termini (Yamada et al. 2016)

Figure 6.3b is an example of the process after the receptor receives an external signal. In the case of rhodopsin, light stimulation causes isomerization of retinal, which changes the intracellular side structure of rhodopsin. In the case of receptors for chemicals in the external environment, molecular binding to the chemicals causes structural changes in the intracellular portion of the receptor protein. These conformational changes allow the receptor to bind to the intracellular G protein, which is a complex of three subunits: α, β, and γ. When the G protein binds to the receptor in the membrane, the G protein dissociates into an α subunit and a βγ subunit. One of them binds to the effector, which determines its function. Furthermore, there are several types of G proteins with different properties, and they perform different functions. In the case of Fig. 6.3b, adenylyl cyclase facilitates the reaction from ATP to cyclic AMP, while protein kinase A amplifies information by binding to AMP. Generally, these phenomena are due to large fluctuations in the three-dimensional structure of the protein.

6.3 Intercellular Communication by Single Transmembrane Proteins

When multicellular organisms emerged, intercellular communication became necessary. To this end, a number of secretory proteins and their corresponding receptors, single transmembrane proteins, were developed. Secretory proteins are a group of proteins whose signal peptides are cleaved and released to the extracellular space. Insulin and growth factors are examples. On the other hand, single transmembrane receptors have a domain on the outside of the cell that binds to a signaling molecule (secreted protein). In addition, it has an inner domain with enzymatic activity for phosphorylation. Both the inner and outer domains are highly water-soluble, and there is only one hydrophobic transmembrane helix that connects them to the membrane, making the protein as a whole a rather low-hydrophobic membrane protein.

This type of receptor is also called an enzymatic receptor because it has enzymatic activity within the cell. The mechanism by which transmembrane receptors transmit information from the outside to the inside of the cell is surprisingly simple. As shown in Fig. 6.4, it forms a complex with two signal molecules and two receptor molecules. As a result, the two receptor molecules are brought into close

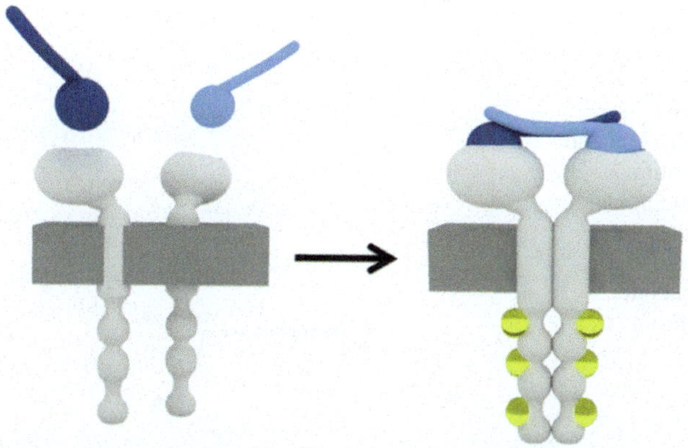

Fig. 6.4 Signal transduction by dimerization of a single transmembrane protein receptor. Once the receptor forms a dimer outside the cell, an intracellular phosphorylation reaction occurs, and subsequent intracellular signaling begins. From Mitaku (2015)

proximity within the cell. Since the interiors of the two receptors have the enzymatic activity of phosphorylation, they phosphorylate each other as they approach. Phosphate has a negative charge and changes the conformation of the protein. This allows them to bind to second messengers and alter the function of the cell.

The number of secreted proteins and single transmembrane receptors appears to be increased in multicellular organisms. This correlation is reasonable since multicellular organisms should require intercellular communication. It may also be related to the exon–intron structure of genes in multicellular organisms. As mentioned above, single transmembrane proteins are less hydrophobic, even though they are membrane proteins. It is statistically difficult to form such an amino acid sequence with large numbers of random mutations. However, it may be easy to incorporate a transmembrane helix into a gene by means of an exon–intron structure.

6.4 Electrical Signals in Neurons

Signal transduction by the nervous system is an electrical signal, which is different from information transduction by chemicals. Most cells have a bias in ion concentrations inside and outside the cell. Intracellularly, potassium ion concentrations are high, and sodium concentrations are low. Conversely, outside the cell, potassium ion concentrations are low, and sodium concentrations are high. This bias in ion concentrations results in a potential difference between inside and outside the cell. The magnitude of this potential is determined by the permeability of each ion through the membrane.

The ion concentration imbalance is formed by membrane proteins (potassium pump and sodium pump) that actively transport potassium and sodium ions. The potassium pump transports potassium from outside to inside the cell, while the

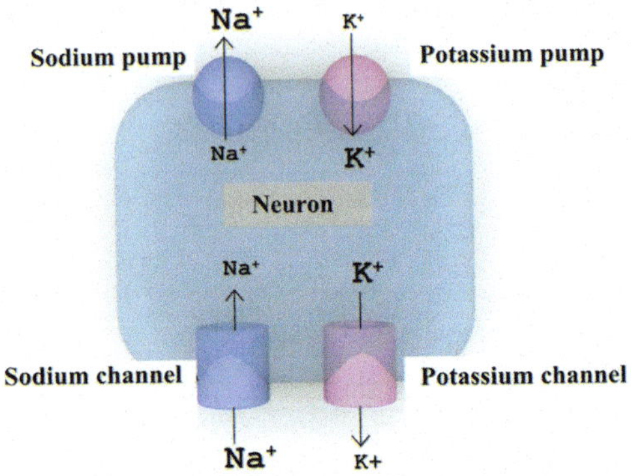

Fig. 6.5 In general, the sodium and potassium pumps result in higher concentrations of intracellular potassium and extracellular sodium. Neurons have sodium and potassium channels, and transient changes in their permeability generate neural signals. From Mitaku (2015)

sodium pump transports sodium from inside to outside the cell using energy such as ATP. There are also proteins that transport potassium and sodium ions simultaneously in opposite directions. This creates an environment for nerve signal transduction (Fig. 6.5).

Potassium and sodium channels, on the other hand, are the main players in membrane potential formation. Although these channels simply allow ions to pass through according to a concentration gradient, their dynamic structure is very cleverly designed in terms of ion selectivity and time course. First, although potassium and sodium ions are difficult to distinguish because of their very similar physicochemical properties, these channels exhibit high selectivity. In addition, these channel proteins have domains called gates, the opening and closing of which can determine the passage of time. Normally, the permeability of potassium ions is slightly higher, resulting in a potassium potential (negative potential with respect to the outside). However, when a signal arrives, the permeability of sodium ions temporarily increases, generating a spike corresponding to the sodium potential. Thus, a signal pulse is generated, which becomes a nerve signal.

When the signal in the nerve axon reaches the next cell, neurotransmitters are released at the synapse. The next cell then has receptors for the neurotransmitter, and the nerve signal is generated again. However, the next cell has input signals from many neurons, and as a result of their computation, a neural signal is generated and more sophisticated information processing takes place. Basically, two types of ion pumps and two types of ion channels allow for very sophisticated information processing. Very complex systems are formed with simple materials, and biological systems are indeed very clever.

6.5 Metabolism by Enzymes

In the first half of this chapter, we discussed how conformational fluctuations can cause proteins to bind and dissociate, thereby enabling signal transduction. Similarly, such conformational fluctuations play an important role in metabolic reactions by

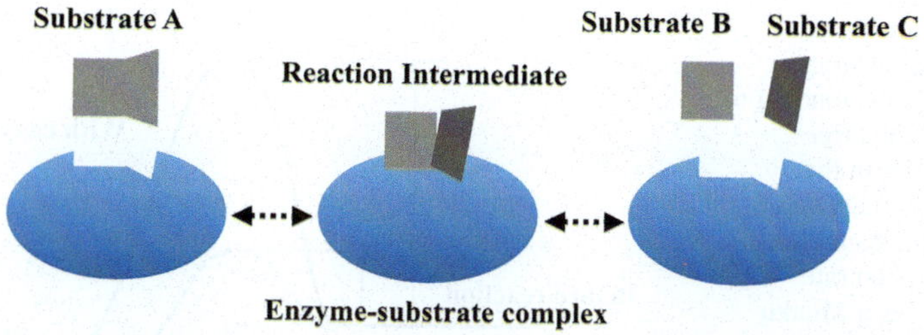

Fig. 6.6 Model of enzyme catalysis. The structure of the substrate molecule bound to the enzyme is distorted. Therefore, the enzyme-substrate complex is prone to change on both the synthesis and degradation sides

enzymes. In both signal transduction and enzymatic metabolic reactions, functions are performed by molecular recognition with other molecules. In the case of enzymes, however, the binding of molecules creates distortions in the steric structure of the metabolite, which facilitates the reaction to degrade or synthesize the metabolite (Sadava et al. 2008) (Fig. 6.5).

Figure 6.6 is a model diagram of enzyme action. The enzyme itself does not change, but it facilitates the reaction. The enzyme binds to substrate A, but it can also bind to its breakdown products, substrates B and C. In other words, during the course of the reaction, the enzyme and substrate combine to form a complex. At this time, the metabolite is easily distorted from the stable substrate A to the degradation products (substrate B and substrate C). In other words, the shape of the complex makes it easy to become substrate A and easy to become degradation products of substrates B and C. If there is more substrate A, the degradation proceeds to substrate B and substrate C. Conversely, if there is more substrate B and substrate C, the synthesis of substrate A proceeds.

Figure 6.7 shows the general relationship between the free energies in each of these metabolite states. Before and after a reaction, there is an activation energy barrier that must be overcome for the reaction to proceed. The reaction rate is not determined by the energy difference before and after the reaction. The reaction rate k is determined by the activation energy E_a at the reaction intermediate, as shown in Eq. (6.1).

$$k = A\exp\left(-\frac{E_a}{\mathrm{RT}}\right) \tag{6.1}$$

At this time, when the metabolite binds to the enzyme, the activation energy is greatly reduced. In other words, when the enzyme binds to the metabolite, the structure of the metabolite changes, the activation energy changes, and the reaction is accelerated. Figure 6.7 models the case of a degradation reaction, but the mechanism of action of the enzyme seems to be almost the same.

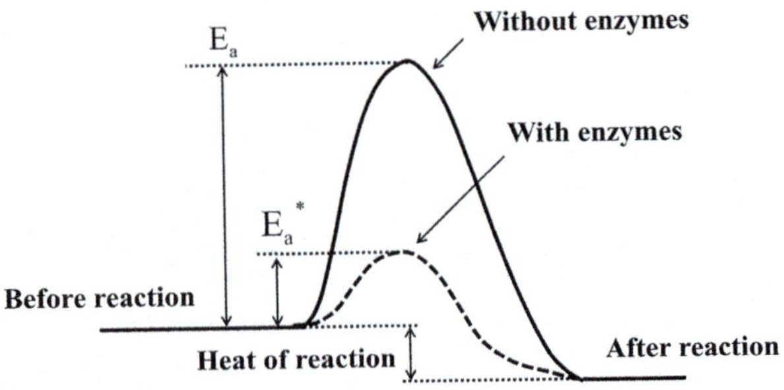

Fig. 6.7 Free energy change of a substrate molecule. Upon binding to the enzyme, the free energy barrier in the reaction intermediate of the substrate decreases, and the reaction rate increases. From Mitaku (2015)

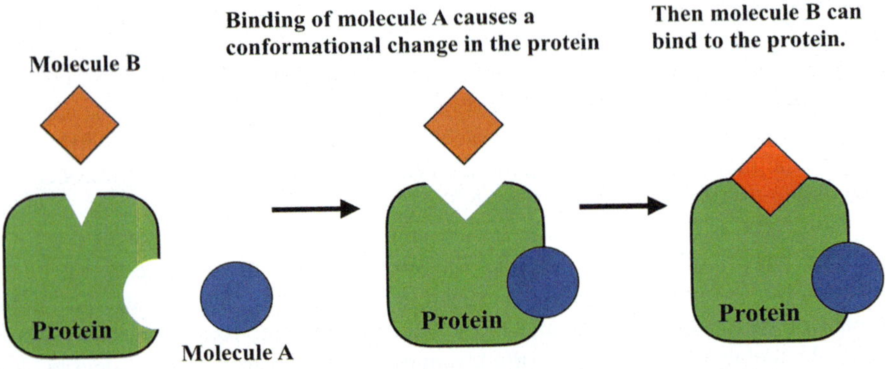

Fig. 6.8 Many proteins have flexible structures. Binding of a regulatory molecule to a site other than the metabolic binding site can alter the original metabolic reaction rate, which is the allosteric effect

6.6 Relationship Between Protein Fluctuations and Function: Allosteric Effect

The advantage of having large fluctuations in proteins is best illustrated by looking at the allosteric effect. A model example of the allosteric effect is shown in Fig. 6.8. In this example, consider a reaction in which a square molecule binds to a protein. This binding does not occur alone, but when the round molecule binds elsewhere in the protein, the binding site of the square molecule opens up, allowing binding to occur.

This effect appears at various places in the biochemical reaction network. Therefore, it is believed to be closely related to the general properties of proteins. That is, proteins are inherently quite fluctuating and have the ability to bind to multiple molecules. When one molecule binds, the binding of another is affected. This is known as the allosteric effect. The allosteric effect itself in individual proteins is a very clever mechanism, but in the context of the genome as a whole, it is a very common phenomenon. The question then arises as to why so many proteins with large fluctuations are produced. We will discuss this issue in Part III.

References

Mitaku S (2015) A modern approach to biological science. Kyoritsu-Shuppan Co., Tokyo (in Japanese)

Sadava D et al (2008) Life, 8th edn. Sinauer Associates, Sunderland, MA

Yamada T, Yamato T, Mitaku S (2016) Forced unfolding mechanism of bacteriorhodopsin as revealed by coarse-grained molecular dynamics. Biophysical J 111:2086–2098

Chapter 7
System Biology and Protein Structure Prediction by Computer

Keywords System biology · Whole genome sequencing · Omics · Homeostasis · Protein structure prediction · Deep learning

In the twenty-first century, whole-genome sequencing has become possible, and the amino acid sequence of all proteins can be revealed by genome sequences. If the functional dynamics of proteins can be written in equations, it should be possible to write all biological reactions in equations. It is based on this idea that the discipline of systems biology was established. However, it is very difficult to completely formulate the dynamics of protein function, even for a single molecule. Therefore, research has been conducted in the direction of understanding the characteristics of system behavior by narrowing down the objectives and eliminating nonessential parameters. Another direction of theoretical research is the prediction of protein tertiary structure. Bottom-up biological research requires knowledge of the function of all proteins. Therefore, the prediction of the tertiary structure of proteins, which is deeply related to their functions, has been attempted for a long time. In recent years, it has become possible to predict structures with high accuracy.

7.1 Attempt to Understand the Whole Organism by Calculation: System Biology

Genome analysis has enabled us to obtain large amounts of data about living organisms. We refer to all DNA sequences of an organism as the genome, all transcripts derived from the genome as the transcriptome, all proteins as the proteome, and all small-molecule metabolites as the metabolome. All these molecules are collectively called omics. Furthermore, the idea of using those large amounts of data to simulate the structure, dynamic behavior, and stability of the entire organism naturally arose. This gave rise to systems biology (Kitano 2002; Kondo et al. 2010).

However, in order to perform simulations, the function and dynamics of each molecule must be known. And even for a single molecule, it is quite difficult to know the

S. Mitaku, R. Sawada, *Evolution Seen from the Phase Diagram of Life*, Evolutionary Studies, https://doi.org/10.1007/978-981-97-0060-8_7

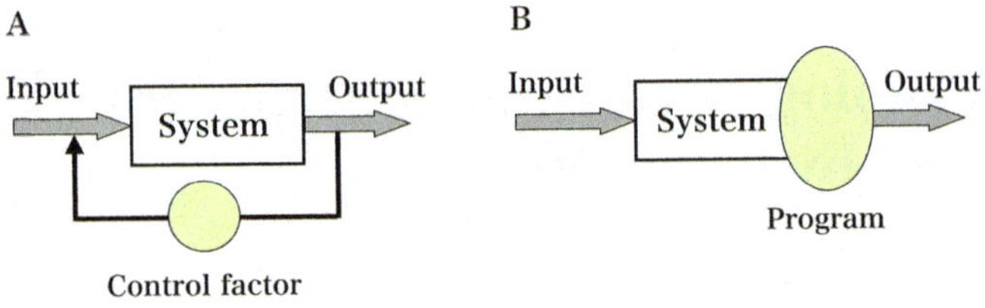

Fig. 7.1 There are two main ways to control a system. They are feedback control (**a**) in which the output is returned to the input, and feedforward control (**b**) in which the behavior is predicted from the output. Each has its advantages and disadvantages

dynamics in detail. In reality, therefore, we have to simplify the simulation by narrowing down the objective and discarding parameters that are not very relevant.

Organisms exist stably in a variety of environments. When the environment changes significantly, they shift to another stable state. This stability of an organism is called homeostasis. The issue of stability can be discussed using the concept of control used in the field of mechanics.

Control methods include feedback and feedforward. As shown in Fig. 7.1, the relationship between inputs and outputs determines the control method. In feedback control, the value of the output is returned to the input. When the negative value of the output is returned to the input, it is called negative feedback. On the other hand, when the positive value of the output is returned to the input, it is called positive feedback.

Often, negative feedback is used to stabilize the system. Conversely, positive feedback is used to control the system to amplify and oscillate it. In biology, negative feedback is often used to stabilize a system. However, because feedback control uses output values for control, there is inevitably a delay in response time.

Therefore, feedforward control is used when speed is needed. For example, consider the task of making a hand robot stand on a bar in a stable manner. If we try this task with feedback control, it is difficult to make the stick stand up stably. This is because the reaction speed is slow. However, if you create a feed-forward control program after many learning cycles, the stick will stand very well. This is because feed-forward control is fast. However, because it relies on a learned program, it may not be able to respond to completely unpredictable inputs.

In the case of biological systems, whether feedback or feedforward, if we can identify the proteins involved in control, we can simulate behavior with fewer parameters.

7.2 Simulation by Mathematical Models

Compared to the complexity of the phenomena exhibited by living organisms, the number of genes contained in the genome is very small. For example, the human genome contains just over 20,000 genes. Although 20,000 genes is a considerable number, the number of human cells is tens of trillions, and it is very strange that all behavior is determined by approximately 20,000 genes. However, when we simulate it in a mathematical model, we can understand why a small number of factors can produce complex behavior.

A cell can be modeled as in Fig. 7.2. Cells secrete factors that activate themselves. In addition, cells are regulated by factors from surrounding cells and secrete factors to control other cells. Thus, the state of the cell is determined by these factors, at which time the various factors diffuse spatially, causing time delays and various phenomena. Input/output factors include hormones, cytokines, ion concentrations, and mechanical stimuli, while cell states include factor secretion, deformation, differentiation, motility, and cell death. In this way, complex phenomena can be caused by a small number of factors.

We will use this model to consider whether the macroscopic properties of an organism can be reproduced by the local inputs and outputs of the cell. A. Turing was the first to tackle this problem. According to his work, macroscopic stationary or traveling waves can be generated by local dynamic reaction fields (Turing 1952). Using this idea, Kondo's team examined the stripe patterns of real animals. The patterns seen on the skin of tropical fish appear to exist statically. However, they are actually formed as dynamic, stationary waves. The observation of changes in the stripe patterns of tropical fish matched well with the simulation by the mathematical model (Kondo and Asai 1995).

In the case of tropical fish stripes, it was just a stationary wave, but there are cases where the behavior of an organism can be explained by a simple traveling wave. For example, in slime molds, individual cells behave like amoebas when food is abundant. However, when food becomes scarce, the cells begin to move cooperatively and unidirectionally. They then form tower-like shapes and begin to move more like slugs. Even these complex cell movements ultimately occur within a single framework with few factors (Weijer 2009).

There is a field in physics called complex systems. It is possible to understand changes in the macroscopic state of an organism, such as cell differentiation, with the idea of complex systems. This is called complex systems biology, which studies the universal properties of living systems (Kaneko 2009). In this theoretical

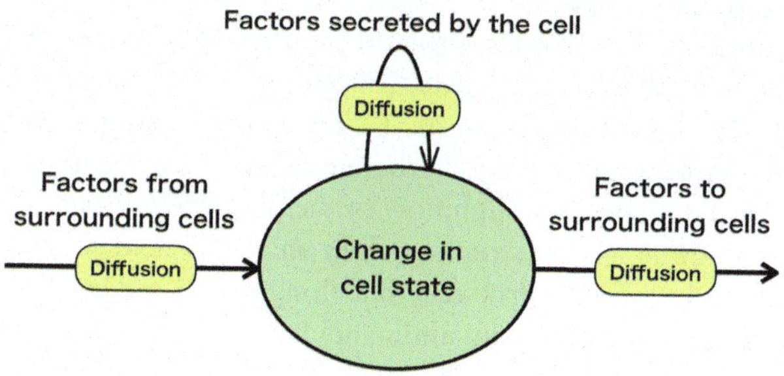

Fig. 7.2 To simulate the behavior of a biological system on a computer, the state of a cell is modeled by the output, diffusion, and input of each factor

framework, a state is represented as a set of many chemical components, and changes in that state are represented as trajectories. When a trajectory converges to a point, it is called a fixed point attractor. When the trajectory oscillates, it is called an oscillating attractor. In addition, it may converge to a strange, irregularly oscillating attractor, called chaos.

The state of a biological system can have multiple attractors and can move from one attractor to another by changing parameters. Once a state is drawn into one attractor, the theory shows that the state is stable, even if there is some noise. Although we will not discuss specific examples here, it is a good general explanation of homeostasis and robustness in living organisms.

Organisms are complex but very stable systems. To understand that the human genome, with approximately 20,000 genes, governs the behavior of trillions of cells, it is also important to think of organisms as systems.

7.3 Protein Structure Prediction and Function

Another way to understand organisms from whole-genome sequences by computers is to predict the three-dimensional structure of all proteins and reveal their functions. This is followed by the elucidation of protein–protein interactions and interactions with metabolites. At present, however, it is difficult to elucidate the function of all proteins by computers. Therefore, the first goal is to elucidate the three-dimensional structure of all proteins that are related to their functions to some extent.

There is an "Anfinsen dogma" (Anfinsen 1973) that states that under appropriate environmental conditions, the three-dimensional structure of a protein is determined solely by its amino acid sequence. If this is so, it should be possible to predict the three-dimensional structure of a protein from the amino acid sequence alone, and many structure prediction methods have been reported. Since 1994, the Competition for Protein Structure Prediction (CASP) has been held. Recently, a new method for predicting protein 3D structure has been developed, and it is now possible to predict the structure with an accuracy of better than 90%. This method (Jumper et al. 2021) is a computer technique based on deep learning. Here, we will look at this method from a physical point of view.

As background for the tertiary structure prediction method AF2, a very large database of amino acid sequences (over 200 million) and protein tertiary structures (approximately 200,000) already exists. Using these large amounts of data, three-dimensional structures have been predicted from amino acid sequences alone. Therefore, we will explain the rough flow of the prediction process.

To fold a three-dimensional structure from an amino acid sequence, both short-range order, such as a helix structure, and long-range order between secondary structures, i.e., the joining of distant amino acids, are needed. Corresponding to this structural feature, both short-range and long-range interactions are at work in terms of physical interactions. There are several physical interactions that maintain protein conformation, including hydrophobic interactions, electrostatic interactions, van der Waals forces, and hydrogen bonds. Van der Waals forces and hydrogen bonding are relatively short-range interactions, while hydrophobic and electrostatic interactions are relatively long-range. However, both contribute to both short- and long-range interactions, and their distance dependence is also very different. This makes a simple physical treatment of protein folding difficult and prevents highly accurate structure prediction.

In addition, enzymes such as chaperones and translocons are essential for protein structure formation. It is not possible to reproduce them exactly on a computer. However, as stated in the Anfinsen dogma, the amino acid sequence contains the information necessary to form the three-dimensional structure of a protein, and the structure formed is physically stable.

Considering the difficulty of such a highly accurate 3D structure prediction method, the AF2 prediction method seems to cleverly incorporate information from both short-range and long-range orders. As shown in Fig. 7.3, AF2 uses two types of amino acid sequence analysis methods. First, for the amino acid sequence of the protein whose 3D structure is to be predicted, a number of similar amino acid sequences are selected from a large database. Then, multiple sequence alignment (MSA) of those similar amino acid sequences is performed. Next, a direct sequence association (DCA) analysis is performed to predict pairs of amino acids that are likely to be in direct contact with each other (Weigt et al. 2009; Morcos et al. 2011). It is known that when a mutation occurs in one amino acid, the amino acids in contact with it also mutate, reducing the stress on the structure. This is called the coevolution of amino acid sequences. DCA is then performed using this coevolution, and AF2 combines these two types of analysis to predict 3D structure.

In fact, both MSA and DCA are thought to contain the effects of both short-range and long-range interactions. Therefore, using the conformation of a similar protein as a template, we can determine where each amino acid is located in the conformation. This allowed us to evaluate the importance of each amino acid in the structure formation. Further updating MSA and DCA using the structure thus obtained seems to improve the accuracy of 3D structure prediction. We believe that structure prediction with an accuracy of over 90% by AF2 has a significant impact not only on structural biology but also on biology as a whole.

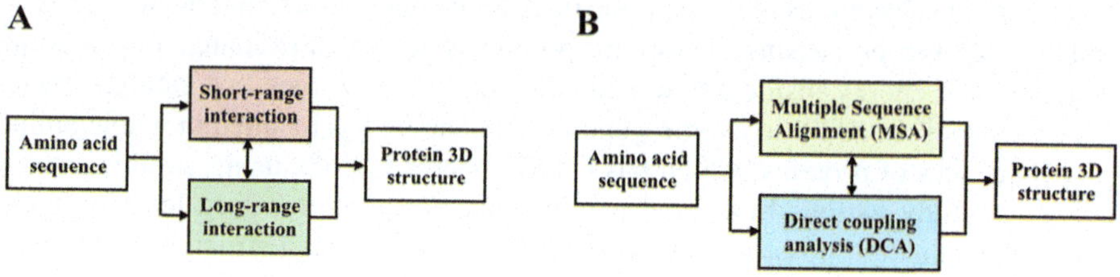

Fig. 7.3 The three-dimensional structure of a protein is formed by a combination of short-range and long-range interactions between atoms (**a**) To predict this with high theoretical accuracy, a method combining multiple sequence alignment (MSA) and direct coupling analysis (DCA) is used (**b**)

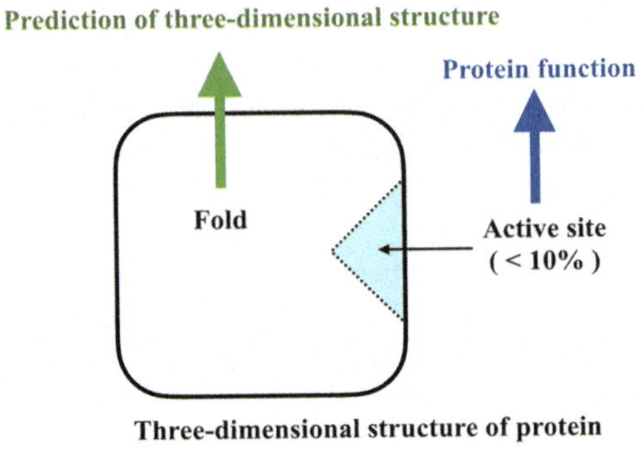

Fig. 7.4 While the amino acids in the active site, which determine the function of the protein, represent only a few percent of the total, the fold, the overall shape of the structure, is determined by the total amino acid sequence. Factors that determine the fold are considered important in discussing the phase diagram of life

However, structure prediction focuses primarily on the three-dimensional structure of proteins, whereas systems biology focuses on function. As mentioned earlier (Chap. 3), it is very important to determine the overall 3D structure of a protein at a somewhat lower resolution fold. In contrast, protein function requires understanding the high-resolution structure of the active site, which is only a few percent of the total amino acids (Fig. 7.4). Thus, even if it becomes possible to predict three-dimensional structures from amino acid sequences, it is difficult to connect them directly to systems biology and to understand life theoretically.

7.4 Understanding Life Through Genome Sequences

The genome sequence is the blueprint for the creation of an organism, and the resulting organism based on the blueprint is an enormous system composed of all proteins. In Part II, we have mainly discussed how proteins are formed and how they interact to form subsystems using a bottom-up approach. In this last section of Part II, we will consider the relationship between Part II and the following Part III.

Important subsystems of the genome include the many molecular devices responsible for genome processing, membrane protein systems that exchange information, materials and energy inside and outside the cell, and enzymes that catalyze metabolic reactions. In addition, whole-genome sequencing has resulted in whole amino acid sequences of proteins, which has led to systems biology using computer technology to study various biological phenomena. Furthermore, deep learning techniques have made it possible to predict the three-dimensional structure of proteins with a high degree of accuracy.

Despite the vast amount of bottom-up research presented in Part II, whole-genome sequencing leaves many major open questions. First, although organisms exhibit a high degree of order, their mechanisms have long remained a mystery. This was already pointed out by Schrödinger and Monod around the middle of the twentieth century (Schroedinger 1944; Monod 1971). Later, it became clear that many random mutations had been introduced into the genome sequence. It was then discovered that, in fact, the genome sequence itself is basically formed by random mutations. Furthermore, studies of molecular evolution have shown that mutations are neutral (Kimura 1968; King and Jukes 1969; Saitou 2009). In other words, a high degree of order in organisms is produced by neutral random mutations. This is a great mystery that has yet to be solved.

Second, it is believed that the accumulation of many mutations in the genome gives rise to new species. However, there is a question as to how many random mutations give rise to a new species, and little conclusion has been reached on this issue. The process by which new species are born is also not well understood. The question of whether new species are born through a gradual accumulation of mutations or through "punctuated equilibrium" is also unresolved (Gould and Eldredge 1993). This question should also be considered in terms of the nature of the whole genome.

These questions cannot be answered directly by a bottom-up approach. We have considered that organisms should not be understood by proteins in three dimensions but by statistical considerations of random mutations on a one-dimensional genome sequence. Then, in Part III, we will discuss how order emerges from random mutations and why mutations are neutral.

References

Anfinsen CB (1973) Principles that govern the folding of protein chains. Science 181:223–230. https://doi.org/10.1126/science.181.4096.223

Gould SJ, Eldredge N (1993) Punctuated equilibrium comes of age. Nature 366:223–227

Jumper J, Evans R, Pritzel A et al (2021) High accurate protein structure prediction with AlphaFold. Nature 596:583–589

Kaneko K (2009) What is life—toward complex life science. The University of Tokyo Press, Inc., Tokyo (in Japanese)

Kimura M (1968) Evolutionary rate at the molecular level. Nature 217:624–626

King JL, Jukes TH (1969) Non-Darwinian evolution: most evolutionary change in proteins may be due to neutral mutations and genetic drift. Science 164:788–798

Kitano H (2002) Systems biology: a brief overview. Science 295:1662–1664

Kondo S, Asai R (1995) A reaction-diffusion wave on the skin of the marine angelfish *Pomacanthus*. Nature 376:765–768

Kondo S, Kitano H, Kneko K, Kuroda S (2010) Introduction to modern biological sciences 8: system biology. Iwanami Shoten Co., Tokyo

Monod J (1971) El hansard et la necessite. Alfred A. Knopf, Inc., Paris

Morcos F, Pagnant A, Lunt B et al (2011) Direct-coupling analysis of residue coevolution captures native contacts across many protein families. Proc Natl Acad Sci U S A 108:E1293–E1301

Saitou N (2009) From selectionism to neutralism: paradigm shift of evolutionary studies. NTT Publishing Co., Tokyo (in Japanese)

Schroedinger E (1944) What is life? The physical aspect of the living cell. Cambridge University Press, Cambridge

Turing AM (1952) The chemical basis of morphogenetic. Philos Trans R Soc Lond B 237(641):37–72

Weigt M, White RA, Szurmant H et al (2009) Identification of direct residue contacts in protein–protein interaction by message passing. Proc Natl Acad Sci U S A 106:67–72

Weijer CJ (2009) Collective cell migration in development. J Cell Sci 122:3215–3223

Part III
Formation of Ordered Structure in Organisms by Random Mutations in Genome Sequences

Chapter 8
Similarities in Order Formation in Matter and Organisms

Keywords Theory of evolution · Natural selection · Repair enzymes · Low entropy state · Phase diagram

One of the major questions about living organisms is how the macroscopic organism as a whole maintains its stability. The elementary processes in the genome of an organism are neutral random mutations, and yet the macroscopic organism maintains its low-entropy state very stably. At first glance, this problem of order formation based on chance seems unsolvable. However, there is a precedent for such a phenomenon in nineteenth-century physics: the establishment of the thermostatistical dynamics of matter. Based on the analogy between matter and living organisms, this chapter discusses the possibility of top-down analysis of genome sequences. To this end, we discussed the two factors that can cause mutations in the genome to be passed on to the next generation: errors in the repair system and natural selection.

8.1 Principles of Physics, Chemistry, and Biology Discovered in the Mid-nineteenth Century

There are periods in the history of science when the discovery of fundamental principles is concentrated. One such period was the mid-nineteenth century, when very important principles were discovered in all areas of physics, chemistry, and biology (Mitaku 2002). Table 8.1 shows some of these discoveries. Although these are principles from independent fields of study, it is clear that they are closely related to each other from the perspective of modern state-of-the-art science.

Let us start with a principle of physics: in 1850, Clausius reported the second law of thermodynamics (Clausius 1850, 1865; Tomonaga 1979). This is the so-called law of increasing entropy. It states that an isolated system gradually becomes disordered. This is a very general principle that applies to all matter, and it established the basic principles of thermodynamics. The second law of thermodynamics completes

S. Mitaku, R. Sawada, *Evolution Seen from the Phase Diagram of Life*, Evolutionary Studies, https://doi.org/10.1007/978-981-97-0060-8_8

Table 8.1 Scientific principles reported from 1850 to 1870

Year	Discovery
1850	Second law of thermodynamics by Clausius
1860	Maxwell distribution of gas molecular motion
1869	Mendeleev's periodic table
1859	Publication of "The Origin of Species" by Darwin
Early 1860s	Pasteur conducted experiments that disproved the spontaneous generation of organisms
1865	Mendel's laws of heredity

the top-down approach to the change of state of matter. Furthermore, it became possible to discuss the stable state of matter through phase diagrams.

In 1860, Maxwell showed that although the motion of gas molecules is random, the distribution of their velocities is constant in equilibrium (Maxwell 1860; Tomonaga 1979). From the Maxwell distribution at equilibrium, we can derive the thermodynamic temperature and pressure. In other words, even for a system consisting of many molecules in random motion, the distribution of random motion is constant at equilibrium. The state of matter can then be characterized by several thermodynamic parameters.

In the field of chemistry, Mendeleev reported the periodic table of elements in 1869 (Mendeleev 1869). This is a simple table showing the periodicity of the properties of the elements. This made it easy to understand the properties of the elements only by their position in the table. This simplicity of the periodic table led to the development of quantum chemistry. It also led to the understanding of biological macromolecules such as proteins and DNA. In this sense, it is the basis of all molecular sciences. The discovery of these principles of physics and chemistry led to the integration of the microscopic and macroscopic states of matter in the mid-nineteenth century.

Let us now look at the principles of biology during the same period. First, in 1859, Darwin published The Origin of Species (Darwin 1859). He spent a long time painstakingly collecting a great deal of biological data and proposed his theory of evolution. His theory of evolution caused great controversy (Saitou 2011). In modern terms, the theory of evolution states that organisms evolve by introducing various mutations into the genome sequence.

In the 1850s, there was still much debate about the spontaneous generation of microorganisms. Then, in the early 1860s, Pasteur conducted an experiment on this issue using a Swann-type flask and proved that organisms do not spontaneously arise. However, the debate about Pasteur's experiment continued for some time. Therefore, he began to describe the methods portion of his paper in detail. This became the standard way of writing papers today so that anyone can reproduce the experiment (Day and Gastel 2006). It is now common sense that without genomes, organisms cannot be created.

Furthermore, in 1865, Mendel reported his laws of heredity (Mendel 1865). There was no debate about this work, and it was forgotten for 30 years until it was rediscovered. The modern way to describe his work is that the function of an organism depends on the individual genes in its genome. This is another essential concept in modern biology.

All three of these studies of organisms are principles about the macroscopic nature of organisms. The study of the microscopic aspects of living organisms developed a century after their studies. Furthermore, half a century later, the possibility of integrating the macroscopic aspects of living organisms (the state of life) with the microscopic aspects (whole-genome sequencing) emerged. This is because whole-genome analysis became possible in the early twenty-first century. Now is the time when we can integrate the micro and macro aspects of living organisms.

8.2 Formation of Ordered Structures by Random Processes: Matter vs. Organisms

The idea of physically understanding the macroscopic aspects of living organisms came in the mid-twentieth century. First, in 1944, E. Schrödinger argued that living organisms as a whole maintain a low-entropy state. Furthermore, J. Monod, in his 1971 book "El hansard et la necessite," discussed the mechanism by which order is formed from random processes. This problem is similar to the one Maxwell tried to elucidate for matter in the mid-nineteenth century. However, the problem of low entropy in living organisms remains unsolved to this day. A closer look at Table 2.1 reveals obvious similarities between matter and living organisms, and we felt it necessary to consider these similarities in more detail.

For matter, there are a number of principles that link the stability of macroscopic states to microscopic random processes (Table 8.2). Here, the leftmost and rightmost entries in Table 8.2 are the same as those in Table 2.1, suggesting that there are similarities between matter and living organisms. For matter, three important principles are presented that link the leftmost and rightmost entries. First, the law of

Table 8.2 Mechanisms of stability in matter and organisms (1). Since both systems form stable states based on random processes, the mechanism of stability in living organisms can be inferred from the mechanism of stability in matter

System	Elementary process	Conservation law	Distribution of units	Parameters for phase diagram	State
Matter(eg. Gas)	Random motion of molecules	Energy conservation	Constant distribution of molecular motion	Intensive parameters in thermodynamics	Three states of matter
Organism	Random mutations in genome sequence	Question1 What is conservation law in genome sequences?	Question 2 What is constant distribution in the biological system?	Question 3 What are the Intensive Parameters for the phase diagram of life?	Ordered state of life

conservation of energy holds at equilibrium for matter. On the other hand, we must consider what conservation laws are established in the genome sequence. This is because in the case of living organisms, the law of conservation of energy does not hold.

Second, molecular motion in matter is completely random, but in equilibrium, the distribution of molecular motion is constant (Maxwell distribution). In other words, it is impossible to predict the motions of individual molecules, but the overall distribution is predictable. The genome sequence, on the other hand, is altered by random mutations. When a mutation is introduced, it causes some change in the protein, but changes in individual proteins are essentially unpredictable. However, it is necessary to examine whether the distribution of the genome's product, the protein, is constant.

Third, in the case of matter, a phase diagram of the state can be drawn with a few parameters (e.g., temperature and pressure). In the case of living organisms, the question is whether a small number of parameters can be used to identify DNA sequences that could serve as the blueprint of an organism from a large number of DNA sequences. In other words, if the boundary between living and inanimate matter can be drawn from the genome sequence with a small number of parameters, it should be a phase diagram of life.

Very recently, astrobiologists have been trying to explore extraterrestrial life for comparison with life on Earth. They are trying to define life more generally (Smith 2016; Cockell 2018). If extraterrestrial life were discovered, its molecular composition and characteristics would be revealed. However, even if extraterrestrials are made up of the same types of molecules as life on Earth, whether they fit into the phase diagram in terrestrial life is another question.

8.3 What Factors Establish Mutations in the Genome Sequence?

It is certain that organisms are designed by their genome sequences and that new species arise as a result of the accumulation of mutations in the genome sequence. The question then arises as to whether changes in the genome sequence are due to intracellular factors, external environmental factors (natural selection), or both. These questions are extremely important in considering the establishment of the genome sequence.

To discuss this issue in more detail, Fig. 8.1 shows the factors that determine an organism's genome in a flowchart. First, many different mutations are introduced into the genome sequence of an organism. However, cells have excellent repair enzymes and mutations are generally repaired correctly (Modrich and Lahue 1996, Sancar et al. 2004). It is also known that with some probability repair will fail. The introduction of mutations into the genome sequence is due to this step. In other words, mutations are introduced by rare repair errors. If the probability of mutation

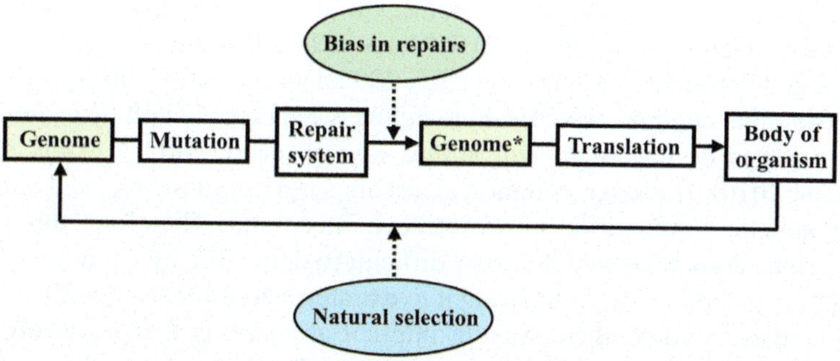

Fig. 8.1 Flowchart of the various factors involved in genome processing. There are two types of factors that determine the genome sequence: intracellular factors (such as repair enzyme groups) and external environmental factors (natural selection). The former intracellular factors can directly determine nucleotide composition but not protein function. In contrast, external environmental factors cannot determine nucleotide composition but will determine functional sequences

is biased, the nucleotide composition may be determined at this stage by this intracellular factor.

However, one more process is required for the mutation to truly remain in the next generation. From the mutated genome sequence, a group of proteins is synthesized, hopefully giving rise to a new organism. The organism is then subjected to natural selection by the surrounding external environment. If the organism survives natural selection, its genome will remain for the next generation. In naive evolutionary theory, it was believed that favorable mutations would remain in the genome because of natural selection. However, it is now established that mutations in evolution are largely neutral (Kimura 1968; King and Jukes 1969; Saitou 2009). Conversely, useless sequences that absolutely cannot design an organism are eliminated in advance.

As shown in the flowchart in Fig. 8.1, there are two factors that determine mutation. The former is an intracellular factor, and the latter is an external factor. Looking at the history of life, organisms have continued for as long as 4 billion years. The question of how these two factors have contributed to the continuation of life is very important. Therefore, throughout Parts III and IV, we will address these questions by analyzing actual genome sequences.

8.4 Analysis of the Genome Sequence with All Genes as Functional Units

Before discussing our analysis, we must review the work that has been done on whole organisms using whole-genome sequences. Since the genome sequences of many organisms became available, questions such as "What is an organism?" and "What is a species?" have been debated. In the course of these discussions, conservation laws for genes and proteins, the units of biological function, were examined (Cleland 2012; Smith 2016; Mariscal and Doolittle 2020; Cohan 2002; Doolittle and Papke 2006; Cohan and Perry 2007; Doolittle 2008; Inkpen et al. 2017).

For example, attempts were made to define the species *E. coli* by examining the common genes of several *E. coli* strains (Welch et al. 2002; Lukjancenko and Wassenaar 2010). However, common genes are surprisingly scarce, and attempts to test for genetic conservation have not been successful. Therefore, this situation raises serious arguments that it is very difficult to define life by common genes and can only be understood by comparing it to extraterrestrial life (Smith 2016; Cockell 2018). In these studies, there was an implicit assumption that organisms have a conservation law for genes. However, it is difficult to define life under this assumption.

The genome sequence is often viewed as a collection of genes or proteins, which means that organisms are discussed as a network of functions. However, the active site of a protein is only approximately 5% of the total amino acid sequence, as shown in Fig. 7.4. In other words, to regard an organism as a network of functions is to ignore more than 90% of the sequence in discussing the organism. It is not common sense, but we have considered that the part of the sequence that is related to function, which is only 5% of the sequence, should rather be regarded as noise to understand the organism as a whole. We should consider all protein folds that are not necessarily related to function.

In the next chapter, we will discuss the distribution of all protein folds in the entire genome sequence. We first analyzed the percentage of membrane proteins among all proteins.

References

Clausius R (1850) Ann Phys Chem Bd 79:S368

Clausius R (1865) Ann Phys Chem Bd 125:S353

Cleland CE (2012) Life without definitions. Synthese 185:125–144

Cockell CS (2018) The equations of life. Atlantic Books

Cohan FM (2002) What are bacterial species? Annu Rev Microbiol 56:457–487

Cohan FM, Perry EB (2007) A systematics for discovering the fundamental units of bacterial diversity. Curr Biol 17:R373–R386

Darwin C (1859) On the origin of species by means of natural selection, or the preservation of favored races in the struggle for life. John Murray, London

Day RA, Gastel B (2006) How to write and publish a scientific paper, 6th edn. Greenwood Press, Santa Barbara

Doolittle WF (2008) Microbial evolution: stalking the wild bacterial species. Curr Biol 18:R565–R567

Doolittle WF, Papke RT (2006) Genomics and the bacterial species problem. Genome Biol 7:116

Inkpen SA, Douglas GM, Brunet TDP et al (2017) The coupling of taxonomy and function in microbiomes. Biol Philos 32:1225–1243

Kimura M (1968) Evolutionary rate at the molecular level. Nature 217:624–626

King JL, Jukes TH (1969) Non-Darwinian evolution: most evolutionary change in proteins may be due to neutral mutations and genetic drift. Science 164:788–798

Lukjancenko O, Wassenaar TM (2010) Comparison of 61 sequenced *Escherichia coli* genome. Microb Ecol 60:708–720

Mariscal C, Doolittle WF (2020) Life and life only: a radical alternative to life definitionism. Synthese 197:2975–2989

Maxwell JC (1860) Illustrations of the dynamical theory of gases. Philos Mag 19:19–32

Mendel G (1865) Experiments on plant hybridization. In: Meetings of the Natural History Society of Brno in Moravia

Mendeleev D (1869) On the correlation between the properties of the elements and their atomic weight. Zhur Russ Khim Obshch 35:60–77. (in Russian)

Mitaku S (2002) Introduction to molecular biology. Iwanami Shoten Co., Tokyo (in Japanese)

Modrich P, Lahue R (1996) Mismatch repair in replication fidelity, genetic recombination, and cancer biology. Annu Rev Biochem 65:lOl-133

Monod J (1971) El hansard et la necessite. Alfred A. Knopf, Inc., Paris

Saitou N (2009) From selectionism to neutralism: paradigm shift of evolutionary studies. NTT Publishing Co., Tokyo (in Japanese)

Saitou N (2011) Introduction to Darwin—Perspectives on modern evolution. Chikuma Shobo Co., Tokyo (in Japanese)

Sancar A et al (2004) Molecular mechanisms of mammalian DNA repair and the DNA damage checkpoints. Annu Rev Biochem 73:39–85

Schroedinger E (1944) What is life? The physical aspect of the living cell. Cambridge University Press, Cambridge

Smith K (2016) Life is hard: countering definitional pessimism concerning the definition of life. Int J Astrobiol 15:277–289

Tomonaga S (1979) What is physics? Iwanami-shoten, Tokyo (in Japanese)

Welch RA, Burland V, Plunkett G III (2002) Extensive mosaic structure revealed by the complete genome sequence of uropathogenic *Escherichia coli*. Proc Natl Acad Sci 99:17020–17024

Chapter 9
Protein Distribution Analysis by a High-Precision Prediction System for Membrane Proteins

Keyword Membrane protein prediction · SOSUI system · Hydrophobicity index · Amphiphilicity index · Proportion of membrane proteins

To determine the proportion of membrane proteins in the whole genome, we developed a membrane protein prediction system, SOSUI. This system used only the physical parameters of each amino acid (hydrophobicity index and amphiphilicity index). This allowed us to predict with an accuracy as high as 95%. The unique feature of this system is that it can make highly accurate predictions even for sequences whose function is unknown. Using this system, we analyzed the entire genomes of many organisms and found that the proportion of membrane proteins in any organism is approximately one-fourth of the total. Furthermore, even in multicellular organisms with splicing, the proportion of membrane proteins was nearly constant. This study was the first step in a series of studies discussing the stability of the entire organism.

9.1 The Meaning of the Proportion of Membrane Proteins in Genome Sequences

Consider the order of protein distribution throughout the genome sequence. Table 8.2 in the previous chapter compares matter and living organisms. In the case of states of matter, when random motions reach equilibrium, the distribution of random motions becomes constant, creating order in the state. If something analogous to this occurs in living organisms, it is possible that many random mutations will reach some equilibrium state and that the distribution of proteins will become constant.

Indeed, whether the distribution of all proteins in the entire genome is constant has never been studied in detail. However, since the 1990s, the homology of protein 3D structures has indicated that there are only approximately 1000 (Chothia 1992). It is certain that many highly resolved 3D structures in the Protein 3D Structure Database (PDB) are classified into their respective folds. Thus, it would be

S. Mitaku, R. Sawada, *Evolution Seen from the Phase Diagram of Life*, Evolutionary Studies, https://doi.org/10.1007/978-981-97-0060-8_9

worthwhile to determine whether the proteins are in a certain distribution in each genome sequence.

To study the distribution of proteins throughout the genome, structural predictions must be made even for proteins whose structures are not yet known. It is very difficult to predict the 3D structure of all proteins from the amino acid sequence alone. Nevertheless, to examine the distribution of folds, it is not necessary to have a high-resolution structure. At this time, we must be able to discriminate folds with high precision, even at low resolution. If it becomes clear that the protein distribution is constant, it could lead to a discussion of the stability of the entire organism.

The argument of 1000 different folds is primarily concerned with soluble proteins. However, to classify all proteins, we must first divide them into two categories: soluble proteins and membrane proteins. Membrane proteins are a group of proteins with transmembrane helices (Branden and Tooze 1998). The transmembrane regions are embedded in a hydrophobic lipid bilayer, while water-soluble proteins are completely soluble in water. In other words, the physical properties of membrane proteins and water-soluble proteins are clearly different in terms of their environment. Therefore, to physically reveal the distribution of proteins, we first discriminate between membrane and water-soluble proteins.

In any case, to reveal the distribution of proteins of all amino acid sequences from the genome sequence, highly accurate prediction based on physicochemical parameters of the amino acid sequence is necessary. A prediction system based on physicochemical parameters can guarantee the same accuracy for all amino acid sequences. Therefore, in the next and subsequent sections, we will focus on highly accurate membrane protein prediction using physicochemical parameters.

9.2 Development of a Membrane Protein Prediction System by Physical Parameters

In the 1980s, little structural data were available for membrane proteins, and several systems were reported that could distinguish membrane proteins based on amino acid sequence alone. Most of these were based on the assumption that the transmembrane helix region was composed of clusters of highly hydrophobic amino acids. However, as the 3D structures of many membrane proteins were analyzed, it became clear that the prediction accuracy based solely on hydrophobic clusters was only 70% at best. Therefore, attempts were made to improve the accuracy by adding empirical rules. However, the problem with the addition of empirical rules was that accuracy for completely unknown amino acid sequences could not be guaranteed. Therefore, we aimed for highly accurate prediction using only physical parameters (Hirokawa et al. 1998).

Lipid molecules are composed of hydrophobic alkyl chains and polar groups. We focused on the fact that lipid molecules aggregate in water to form lipid bilayers. This implies that amphiphilic portions should exist at both ends of the transmembrane helix in membrane proteins as well. In fact, some individual amino acids have amphiphilic side chains. Here, amphiphilic side chains can be defined as having a polar group (charge or electric dipole) at the end and flexible branches connecting the polar group to the main chain (Hirokawa et al. 1998; Mitaku et al. 2002). Amino acids with such side chains include positively charged lysine, arginine, and histidine, negatively charged glutamic acid, and electrically neutral glutamine, tyrosine, and tryptophan. In contrast, aspartic acid, asparagine, serine, and threonine were excluded from the amphiphilic amino acids because they have polar groups but no branching freedom.

The hydrophobicity index (Kyte and Doolittle 1982) and amphiphilicity index (Hirokawa et al. 1998; Mitaku et al. 2002; Tsuji and Mitaku 2004) of amino acids are shown in Table 9.1. The amphiphilicity index was calculated based on the surface area of the polar groups and the branches connecting the main chains. This is because the strength of the attraction between hydrophobic molecules (hydrophobic effect) is generally proportional to the surface area of the molecules. As shown in Table 9.1, we divided amphiphilic amino acids into two types. The reason is that the distribution of the more polar amino acids (lysine, arginine, histidine, glutamic acid, glutamine) and the less polar amino acids (tyrosine, tryptophan) was slightly different in relation to the membrane surface. In any case, it is important to note that both the hydrophobicity and amphiphilicity indices have clear physical significance.

For many membrane proteins, we examined the distribution of hydrophobicity and amphiphilicity indices near the transmembrane region. The results are shown schematically in Fig. 9.1 (Tsuji and Mitaku 2004). Here, we calculated the average index of a segment consisting of several residues in the sequence and plotted it at the center of the segment. The peak of the hydrophobicity index of the segment is located approximately in the center of the transmembrane helix region. The peaks of the amphiphilic indices are located on both sides of the transmembrane region. The distribution trends of these indices were used to construct the SOSUI system to predict transmembrane helices.

The algorithm is so simple that our membrane protein prediction system, SOSUI, was suitable for operation on the Internet (Hirokawa et al. 1998). The system is very

Table 9.1 Hydrophobicity index and amphiphilicity index of amino acids

Amino acid	<H>	<A>	<A′>	Amino acid	<H>	<A>	<A′>
Lysine (K)	−3.9	3.7	0	Threonine (T)	−0.7	0	0
Arginine (R)	−4.5	2.5	0	Proline (P)	−1.6	0	0
Histidine (H)	−3.2	1.5	0	Glycine (G)	−0.4	0	0
Glutamic acid (E)	−3.5	1.3	0	Alanine (A)	1.8	0	0
Glutamine (Q)	−3.5	1.3	0	Methionine (M)	1.9	0	0
Aspartic acid (D)	−3.5	0	0	Cysteine (C)	2.5	0	0
Asparagine (N)	−3.5	0	0	Phenylalanine (F)	2.8	0	0
Tryptophan (W)	−0.9	0	6.9	Leucine (L)	3.8	0	0
Tyrosine (Y)	−1.3	0	5.1	Valine (V)	4.2	0	0
Serine (S)	−0.8	0	0	Isoleucine (I)	4.5	0	0

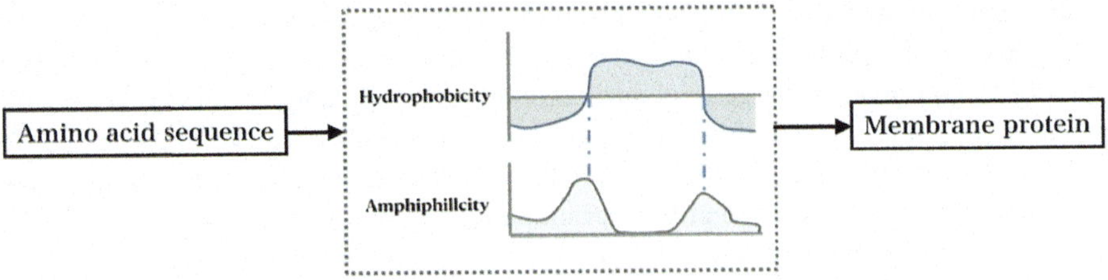

Fig. 9.1 Distribution of hydrophobicity and amphiphilicity of amino acid sequences in a transmembrane helix. The membrane protein prediction system SOSUI uses this distribution of amino acid sequence properties as its algorithm. Adapted from Mitaku et al. (2002)

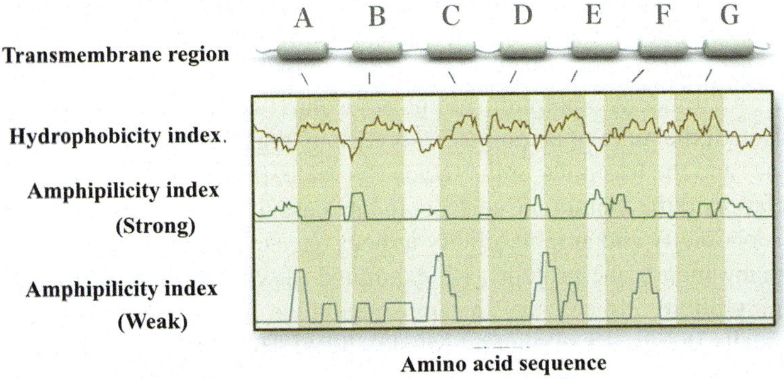

Fig. 9.2 Seven transmembrane helices in the amino acid sequence of bacteriorhodopsin predicted by the membrane protein prediction system SOSUI. The hydrophobic index peak is in the center of the transmembrane helix, flanked by amphiphilic index peaks. From Mitaku et al. (2002)

fast, giving an answer in approximately 1 s after the amino acid sequence is entered into the window. Figure 9.2 shows an example of the analysis of bacteriorhodopsin, a membrane protein with seven transmembrane helices (Mitaku 2015). The graph shows the distribution of the hydrophobicity index and two amphiphilic indices. The horizontal axis represents the position of the amino acid sequence, and the vertical axis represents the window average of the index values. Colored areas represent experimentally determined transmembrane helix regions. Overall, the hydrophobicity index is higher near the center of the transmembrane domain, while the amphiphilicity index peaks at both ends.

The transmembrane helix of bacteriorhodopsin is viewed from the amino terminus. Helices A and B are typical transmembrane helices, with prominent hydrophobic clusters and amphiphilic clusters at both ends. Helices C and E are less hydrophobic at the amino terminus, but this is compensated for by the higher amphiphilicity. Helices F and G have a less hydrophobic valley near the center of the transmembrane domain where functionally important polar amino acids are located. These transmembrane helices are thought to be supported by the structure of the already synthesized transmembrane helices. In addition, the amphiphilic amino acids at both ends of the helix actually seem to contribute significantly to the stability of the transmembrane helix (Mitaku et al. 2002; Tsuji and Mitaku 2004).

The high prediction accuracy and fast analysis speed of the SOSUI system have been highly evaluated. Looking at the number of citations of papers, the cumulative number of citations exceeds 1500 in 2023, and the number of countries and regions accessing the system exceeds 100. This system is widely used and indirectly demonstrates the excellence of the system. The following studies were used in the SOSUI system, which is built on only two physical properties of amino acids: hydrophobicity and amphiphilicity (Hirokawa et al. 1998; Mitaku and Hirokawa 1999; Mitaku et al. 2002; Mitaku 2015).

9.3 The Proportion of Membrane Proteins in the Genome Sequence Is Almost Constant in All Organisms

As discussed in Chap. 8, if the distribution of proteins produced by random mutations is constant, a phase diagram of life with few parameters is likely to be obtained. Thus, using SOSUI, a membrane protein prediction system developed solely on the physical parameters of amino acid sequences, we examined the fraction of membrane proteins in the genome (Mitaku 2013).

The results are shown in Figs. 9.3 and 9.4. Figure 9.3 plots 2551 prokaryotes and 113 eukaryotes. The horizontal axis represents the number of genes in the genome, and the vertical axis represents the percentage of membrane proteins. The fraction of membrane proteins was nearly constant for both prokaryotes and eukaryotes,

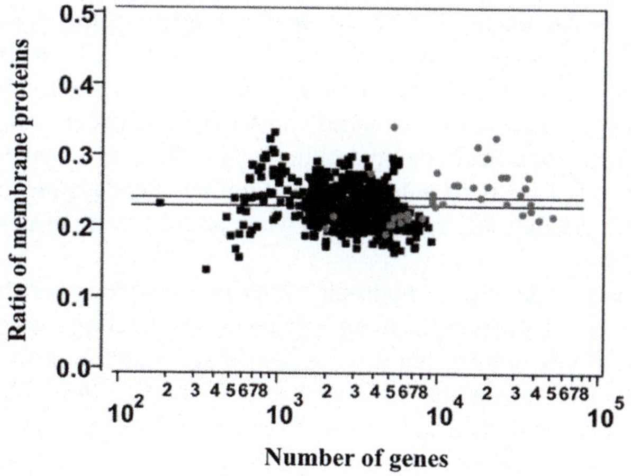

Fig. 9.3 Percentage of membrane proteins in the whole-genome sequences of many organisms predicted by the membrane protein prediction system SOSUI. Black circles are prokaryotes, and gray circles are eukaryotes. Although there is some variation, the average percentage of membrane proteins in both prokaryotes and eukaryotes is approximately 23%. From Mitaku and Sawada (2016)

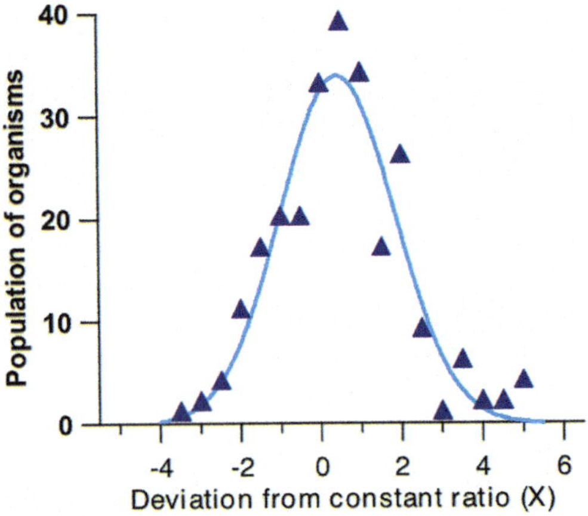

Fig. 9.4 Deviation from the average percentage of membrane proteins shows a Gaussian distribution, indicating that the formation of membrane proteins is determined by random processes. From Sawada et al. (2007)

with an average value of approximately 1/4 (Mitaku and Sawada 2016). Furthermore, the fraction of membrane proteins showed a Gaussian distribution (Fig. 9.4), strongly suggesting that membrane proteins are formed through random processes (Sawada et al. 2007).

Figure 9.5 shows the distribution of genome sizes for many organisms. The genome size of eukaryotes is considerably larger than that of prokaryotes. However, as Fig. 9.3 shows, the proportion of membrane proteins in eukaryotes and prokaryotes is almost the same. In other words, even as prokaryotes evolved into eukaryotes and their genomes grew larger, membrane proteins increased at the same rate. Although eukaryotes and prokaryotes differ in many respects, such as cell shape and size, the proportion of membrane proteins remains almost the same.

We also examined differences in the average hydrophobicity index of all proteins in the genome for various organisms. Figure 9.6 plots the average hydrophobicity of amino acids in all genomes (Mitaku 2015). Surprisingly, the mean hydrophobicity indices were quite different between eukaryotes and prokaryotes. In prokaryotes, the hydrophobicity index was generally distributed between 0.0 and −0.2. In eukaryotes, on the other hand, the hydrophobicity index was distributed at approximately −0.3 to −0.4 for both unicellular and multicellular organisms. This means that prokaryotic proteins are hydrophobic, whereas eukaryotic proteins are systematically hydrophilic.

The systematic reduction of hydrophobicity in eukaryotes is rather rational. In eukaryotes, especially in multicellular organisms, intercellular communication is very important. For example, hormone receptors are single-transmembrane membrane proteins. This means that the water-soluble domains inside and outside the cell are very large, and such membrane proteins are less hydrophobic overall. In addition, eukaryotes have well-developed intracellular organelles, with a wide variety of organelles. Signal peptides are then used to transport proteins to these intracellular organelles. Such signal peptides are generally composed of segments containing many hydrophilic amino acids. This tends to make the overall protein less hydrophobic. Thus, eukaryotes are less hydrophobic overall, but the proportion of membrane proteins remains largely unchanged (Mitaku and Sawada 2016).

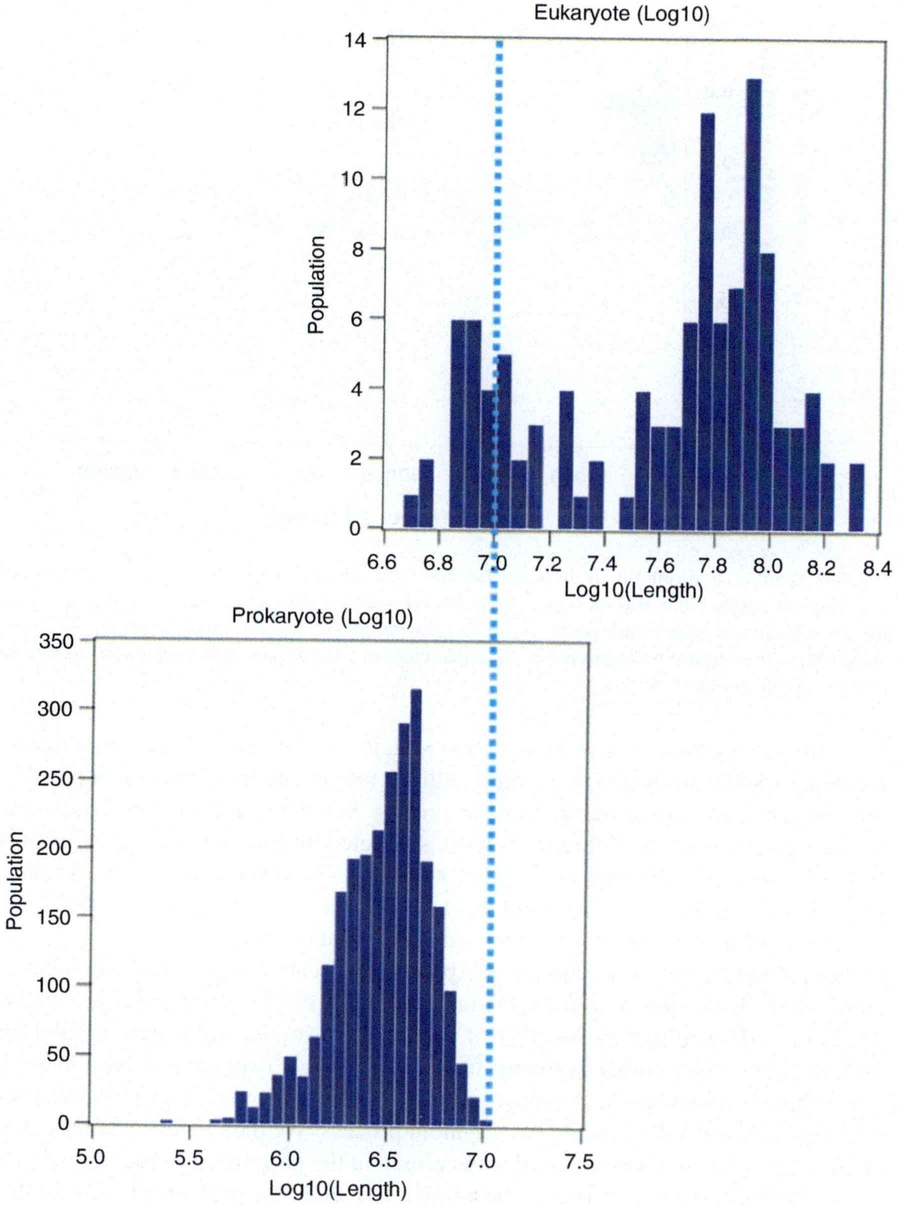

Fig. 9.5 Size distribution of coding regions in prokaryotic and eukaryotic genomes. In terms of the number of nucleotides, the range is $10^{5.7}$–10^{7} for prokaryotes and $10^{6.7}$–$10^{8.2}$ for eukaryotes. In eukaryotes, there are two peaks corresponding to unicellular and multicellular organisms

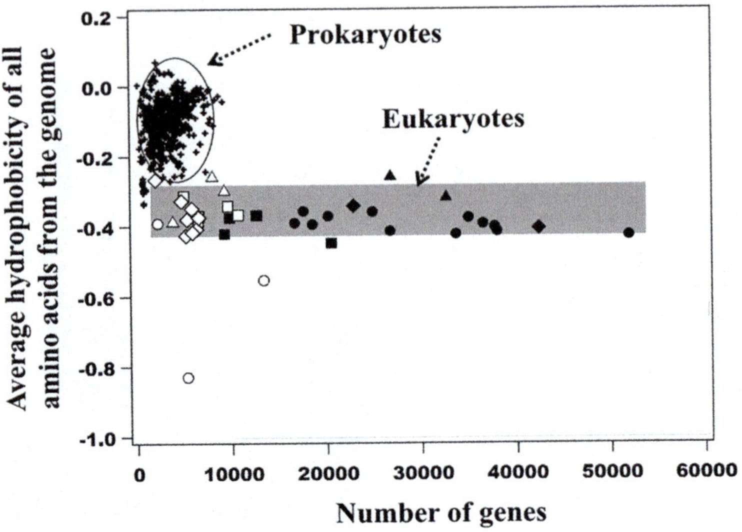

Fig. 9.6 Average hydrophobicity indices obtained from all coding regions of various organisms. The mean hydrophobicity differs significantly between prokaryotes and eukaryotes. Nevertheless, the percentages of membrane proteins are almost the same. In other words, prokaryotes and eukaryotes differ in the contribution of hydrophobic and amphiphilic amino acids in membrane proteins. From Mitaku (2015)

In higher organisms, a process called splicing occurs during the formation of messenger RNA from genes. In multicellular organisms in particular, more than 80% of genes undergo splicing, as shown in Fig. 9.7. The question then arises as to how the proportion of membrane proteins is affected by splicing. As Fig. 9.7 shows, splicing does occur in some unicellular organisms. However, here, we analyzed all genes from genomes in animals and plants to determine the effect of splicing on the proportion of membrane proteins (Sawada and Mitaku 2011).

We plotted all genes in animals and plants on a scatter plot with the number of exons on the horizontal axis and the number of genes on the vertical axis in Fig. 9.8. The results showed that the number of genes had a single exponential distribution with respect to the number of exons. In general, a single exponential distribution is a distribution found in random processes. Thus, it was strongly suggested that the number of exons is determined by a random process. We then investigated the possibility that splicing processes had some effect on the proportion of membrane proteins. The results were surprising: the number of membrane proteins plotted parallel to the single-exponential distribution of total proteins. Since this is a semilogarithmic graph, it shows that the proportion of membrane proteins is approximately one-fourth, regardless of the number of exons (Fig. 9.8). In other words, the splicing process had little effect on the fraction of membrane proteins (Sawada and Mitaku 2011). Although data are not shown here, when the genome was composed of multiple chromosomes, the proportion of membrane proteins was approximately 1/4 for each chromosome.

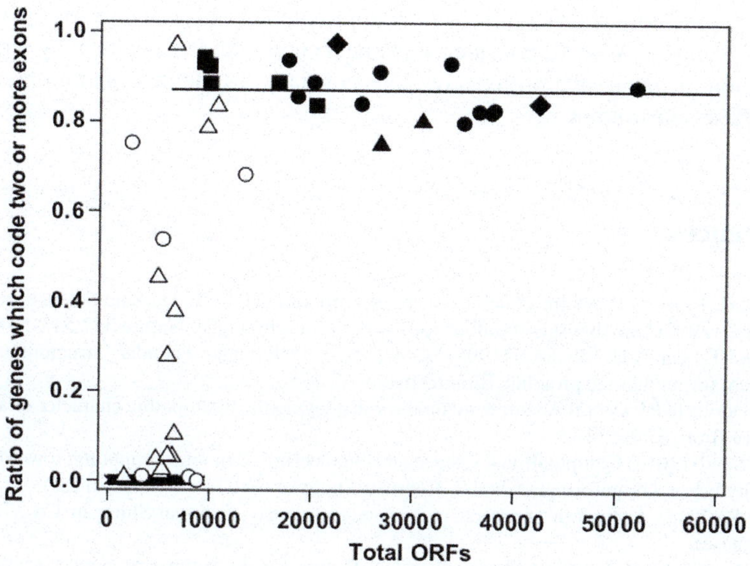

Fig. 9.7 Percentage of spliced genes to total number of genes in various organism genomes. In prokaryotes, no genes are spliced, but in animals and plants, more than 80% of genes are spliced. In fungi and protists, some genes are not spliced, while in some cases, more than 80% of genes are spliced. This correlates well with the process of evolution from unicellular to multicellular organisms. From Sawada and Mitaku (2011)

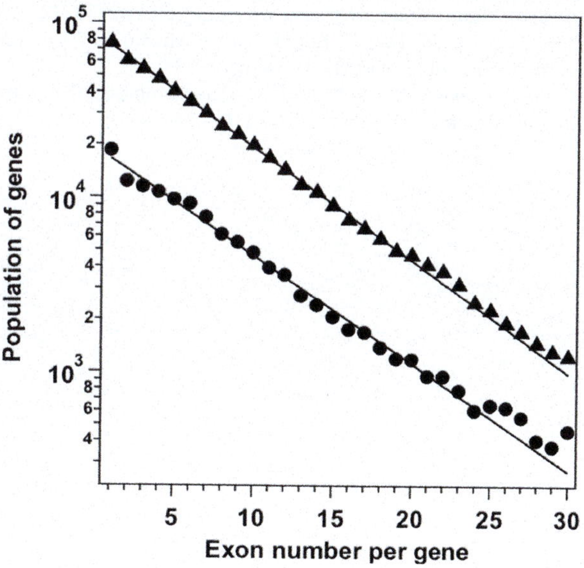

Fig. 9.8 Plotting the number of genes as a function of the number of exons shows a single exponential function. This indicates that exon formation is due to a random process. When the number of membrane proteins was plotted on the same graph, a straight line parallel to all proteins was obtained. This means that for each exon number, the percentage of membrane proteins is constant, approximately 1/4. From Sawada and Mitaku (2011)

Thus, a membrane protein fraction of approximately 1/4 is a universal parameter for all organisms. We will consider in the next chapter the mechanism by which the proportion of membrane proteins remains constant despite this random process of single base substitution.

References

Branden C, Tooze J (1998) Introduction to protein structure, 2nd edn. Garland Science, New York

Chothia C (1992) One thousand families for the molecular biologist. Nature 357:543–544

Hirokawa T, Seah B-C, Mitaku S (1998) SOSUI: classification and secondary structure prediction system for membrane proteins. Bioinformatics 14:378

Kyte J, Doolittle RF (1982) A simple method for displaying the hydropathic character of a protein. J Mol Biol 157:105–132

Mitaku S (ed) (2013) Computational science and engineering, Genome computing; toward genome reality beyond bioinformatics, vol 7. Kyoritsu-Shuppan Co., Tokyo (in Japanese)

Mitaku S (2015) A modern approach to biological science. Kyoritsu-Shuppan Co., Tokyo (in Japanese)

Mitaku S, Hirokawa T (1999) Physicochemical factors for discriminating between soluble and membrane proteins: hydrophobicity of helical segments and protein length. Protein Eng 12:953–957

Mitaku S, Sawada R (2016) What parameters characterize "life"? Biophys Physicobiol., Special Issue " Memorial Issue for Prof. Nobuhiko Saitô" 13:305–310

Mitaku S, Hirokawa T, Tsuji T (2002) Amphiphilicity index of polar amino acids as an aid in the characterization of amino acid preference at membrane-water interfaces. Bioinformatics 18:608–616

Sawada R, Mitaku S (2011) How are exons encoding transmembrane sequences distributed in the exon–intron structure of genes? Genes Cells 16:115–121

Sawada R, Ke R, Tsuji T, Sonoyama M, Mitaku S (2007) Ratio of membrane proteins in total proteomes of prokaryota. Biophysics 3:37–45

Tsuji T, Mitaku S (2004) Features of transmembrane helices useful for membrane protein prediction. Chem-Bio Informatics J 4:110–120

Chapter 10
Changes in the Proportion of Membrane Proteins by Mutation Simulation

Keywords Mutation simulation · Proportion of membrane proteins · Gaussian distribution · Single exponential distribution · Conservation of nucleotide compositions

We Performed mutation simulations on the whole-genome sequence to determine how random mutations affect the proportion of membrane proteins. We performed two types of mutation simulations: one at the amino acid level with fixed amino acid composition and one at the DNA level with fixed nucleotide composition. In both cases, the distribution of membrane proteins matched the actual values only when the actual composition was fixed. In other words, if the nucleotide composition was the same as the actual genome, the amino acid composition would be the actual value even with random mutations, and moreover, the proportion of membrane proteins would be the same. The results of the mutation simulation naturally led to a mechanism by which random mutations determine the distribution of proteins.

10.1 The Proportion of Membrane Proteins Is Constant Even in Mutation Simulations

Evolutionary processes are the result of experiments conducted by nature. However, evolutionary processes such as the birth and extinction of species are one-time experiments that cannot be reproduced under different conditions. In contrast, it is possible to generate various genome sequences on a computer. While it is not possible to discuss the function of all proteins from the sequences, it is possible to discuss the distribution of proteins with a highly accurate protein prediction system. In fact, in the previous chapter, we were able to determine the percentage of membrane proteins in the entire genome. Using SOSUI, a membrane protein prediction system that uses only physical parameters, it is possible to evaluate the percentage of membrane proteins in any given sequence by mutation simulation.

We therefore performed two types of computer simulations to investigate the effect of random mutations on the proportion of membrane proteins (Mitaku 2015). One is a mutation simulation with fixed amino acid composition, and the

S. Mitaku, R. Sawada, *Evolution Seen from the Phase Diagram of Life*, Evolutionary Studies, https://doi.org/10.1007/978-981-97-0060-8_10

other is a mutation simulation with fixed nucleotide composition. In both cases, we compared the actual composition with a completely random composition to determine the effect of composition on the proportion of membrane proteins. The flowchart of the simulation (Fig. 10.1) shows the case of nucleotide mutations (Sawada and Mitaku 2012). The simulation for the case of introducing mutations in the amino acid sequence is basically the same as the flowchart (Sawada et al. 2007). In both cases, the SOSUI system was used to determine the percentage of membrane proteins. One step of the simulation was also performed by introducing mutations at a rate of 1 mutation per 100 codons or 100 amino acids.

Figure 10.2 shows a mutation simulation when the amino acid composition is fixed to a completely random composition (1/20). The horizontal axis in Fig. 10.2a represents the number of steps in the simulation, which corresponds to the evolution time. The vertical axis represents the number of membrane proteins. In this case, the fraction of membrane proteins drops sharply from approximately 1/4 in the real genome to approximately 0.1. This is clearly far from realistic values. A similar computer experiment was performed on the genome sequences of many species, and the values at the plateau are plotted in Fig. 10.2b. The results show that the number of membrane proteins is quite different from reality, being reduced to approximately half of the realistic case. This simulation shows that there is a good correlation between the amino acid composition and the percentage of membrane proteins, indicating that the percentage of membrane proteins is determined by the composition.

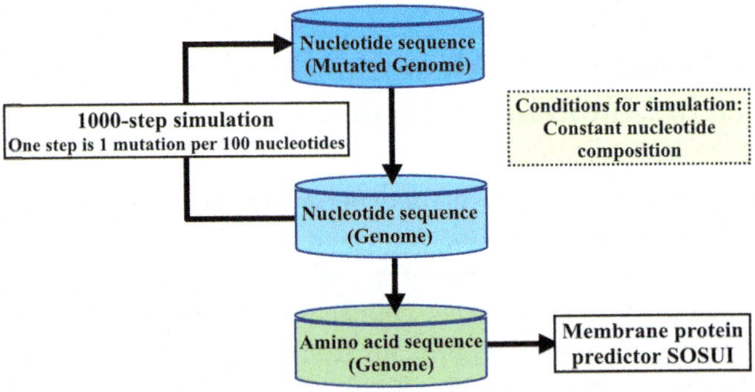

Fig. 10.1 The membrane protein prediction system SOSUI was used to simulate mutations in the genome sequence. Two types of simulations were performed: one at the amino acid level and the other at the DNA level. This figure shows a flowchart of the DNA-level mutation simulation. In the simulations, mutations were introduced randomly while keeping the nucleotide composition constant. In one step of the simulation, mutations were introduced randomly at a rate of one base per 100 bases. The DNA sequence was then converted to an amino acid sequence at each step, and the percentage of membrane proteins in the total genome sequence was calculated using the SOSUI system. Simulations were performed up to 1000 steps. The simulation at the amino acid level followed the same process as that at the DNA level, except that mutations were introduced directly into the amino acids

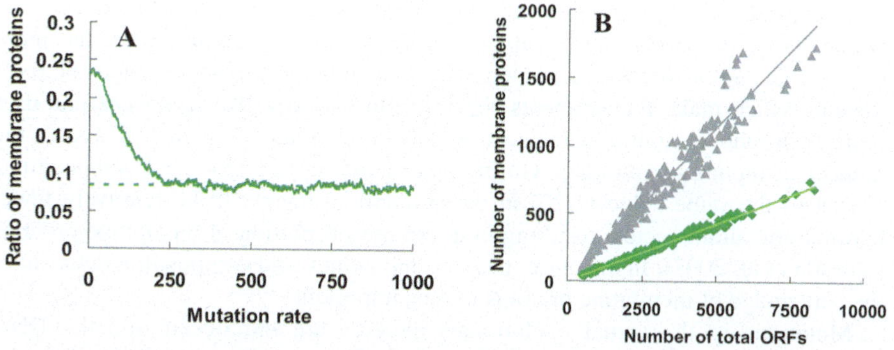

Fig. 10.2 Results of mutation simulation at the amino acid level, where the fraction of 20 amino acids was fixed to the same value of 0.05. (**a**) The proportion of membrane proteins reached equilibrium at 200 steps. (**b**) At equilibrium, the number of membrane proteins was found to be significantly reduced compared to the actual number of membrane proteins. The decrease in the number of membrane proteins was thought to be due to a decrease in hydrophobic amino acids. From Sawada et al. (2007)

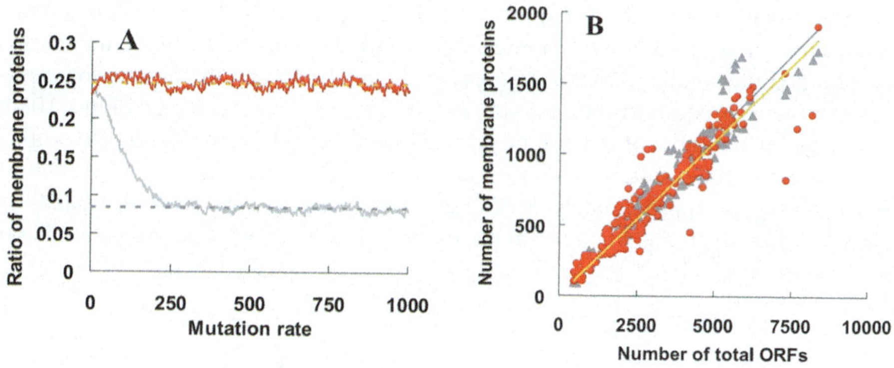

Fig. 10.3 Results of mutation simulation at the amino acid level with amino acid composition fixed to the same value as the actual genome. (**a**) The ratio of membrane proteins was in equilibrium from the beginning. (**b**) The overall ratio of membrane proteins was the same as that of the actual organism. This means that the ratio of membrane proteins is independent of the function of each gene, suggesting that it is determined solely by amino acid composition. From Sawada et al. (2007)

On the other hand, a simulation in which the amino acid composition was fixed to the actual genomic values is shown in Fig. 10.3a. In this case, the ratio of membrane proteins changed little with simulation time. Simulation results for the genome sequences of many other species also showed little change in the ratio of membrane proteins. In addition, the proportion of membrane proteins had the same Gaussian distribution as for the actual genome (not shown in the figure). In other words, the results showed that given the same amino acid composition as the actual genome, the proportion of membrane proteins was almost the same as in reality, even in the simulation experiments with random mutations.

The results of this simulation can be explained as follows. According to Fig. 3.6, which shows the amino acid composition of the *E. coli* genome, there are more hydrophobic amino acids, such as leucine and alanine, and fewer cysteine and tryptophan, which make the segments rigid. In other words, the actual genome uses more hydrophobic amino acids, such as leucine and alanine, which increases the number of membrane proteins. On the other hand, if all amino acid compositions are set to the same value (1/20) in the simulation, there will be relatively fewer hydrophobic amino acids, resulting in a reduced proportion of membrane proteins (Sawada et al. 2007). In any case, the fact that amino acid composition determines the percentage of membrane proteins is very important.

Mutations in the actual evolutionary process are introduced into the DNA sequence. Therefore, it is desirable to introduce mutations into the DNA sequence in the simulation as well. Therefore, we next performed simulations for two types of nucleotide compositions. In the first, only the GC content of the genomic sequence was fixed to realistic values (Fig. 10.4). In the second simulation, we fixed the nucleotide composition at each codon position to realistic values (Fig. 10.5) (Sawada and Mitaku 2012).

Looking at the time course of the simulations in Figs. 10.4a and 10.5a, we can see that a plateau is reached at approximately 200 steps in both cases. This implies that as mutations accumulate, a conversion reaction occurs between the membrane protein and the soluble protein. Figures 10.4b and 10.5b plot the number of membrane proteins in equilibrium versus the number of genes in the genome. Figure 10.4b shows that the number of membrane proteins varies greatly with GC content, with a GC content of 0.7 resulting in almost zero membrane proteins and a GC content of 0.3 resulting in approximately 50% membrane proteins. On the other hand, as shown in Figure 10.5b, when the mutation probability at each codon position is set

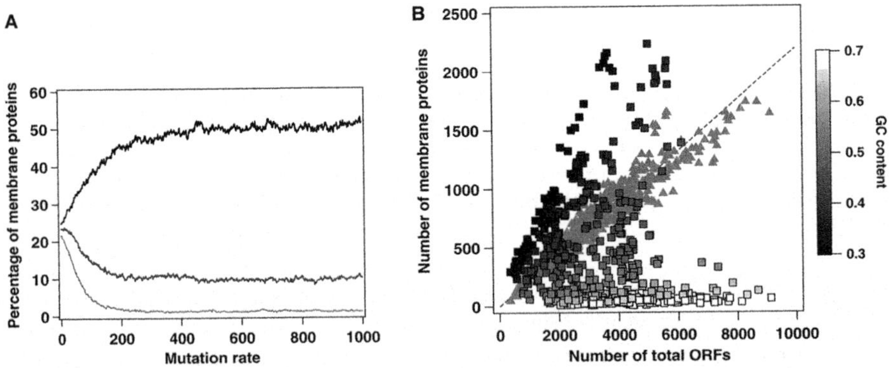

Fig. 10.4 Results of DNA-level mutation simulation performed with fixed GC content. (**a**) The fraction of membrane proteins reached equilibrium in approximately 200 steps, and the value at equilibrium varied greatly with GC content. (**b**) Plotting the number of membrane proteins that reached equilibrium shows that the percentage of membrane proteins is very low when the GC content is high, and conversely, the percentage of membrane proteins is very high when the GC content is low, compared to the actual case. From Sawada and Mitaku (2012)

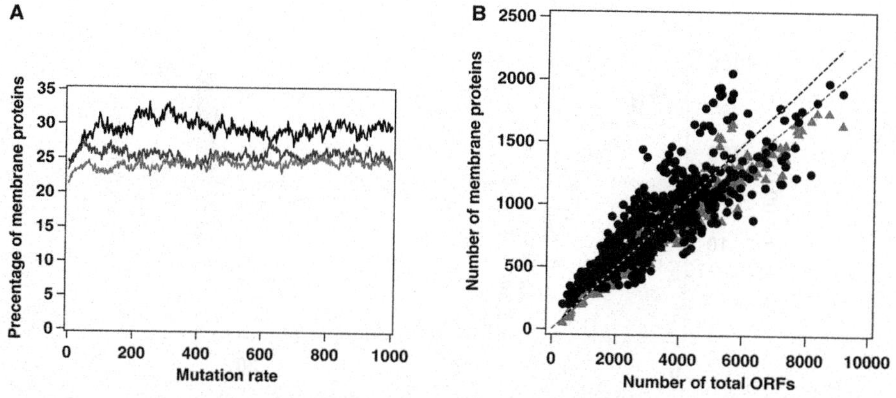

Fig. 10.5 DNA-level mutation simulations with the nucleotide composition at each codon position fixed to that of the actual genome. (**a**) The proportion of membrane proteins was nearly in equilibrium from the beginning. (**b**) Comparing the results of the simulation with those of a real organism (gray marks), the percentages of membrane proteins were almost identical. This suggests that the fraction of membrane proteins is determined by nucleotide composition. From Sawada and Mitaku (2012)

to the actual nucleotide composition, the percentage of membrane proteins does not change much. Each point in Fig. 10.5b is the result of an independent simulation of a different organism, showing that the percentage of membrane proteins is close to the actual value due to random mutations (Sawada and Mitaku 2012).

To understand the relationship between the amino acid level simulation and the DNA level simulation, we used the genome sequence of *E. coli* to examine the relationship between nucleotide composition and amino acid composition. We set the composition at each letter position of the codon to the *E. coli* genome and simulated random mutations. We then translated each step of the simulation using a codon table and examined the amino acid composition. In fact, the values were in good agreement with Fig. 3.6 (Mitaku 2013). In other words, it is natural that the results of the nucleic acid level simulation and the amino acid level simulation should be the same (Figs. 10.2, 10.3, 10.4 and 10.5).

Based on the discussion thus far, there is no relationship between the percentage of membrane proteins in the genome as a whole and the function of individual proteins. On the other hand, it is known empirically that the number of transmembrane helices is strongly related to the function of membrane proteins. For example, many receptor proteins are seven transmembrane proteins. Many transport proteins are six transmembrane proteins. In other words, the number of transmembrane helices seems to increase or decrease depending on function.

Therefore, we performed mutation simulations while ignoring function to see how the number of transmembrane helices changes. If the number of transmembrane helices changes significantly in the mutation simulation, it means that the function of the membrane protein is strongly related to the number of transmembrane helices. Therefore, we performed mutation simulations on the *E. coli* genome

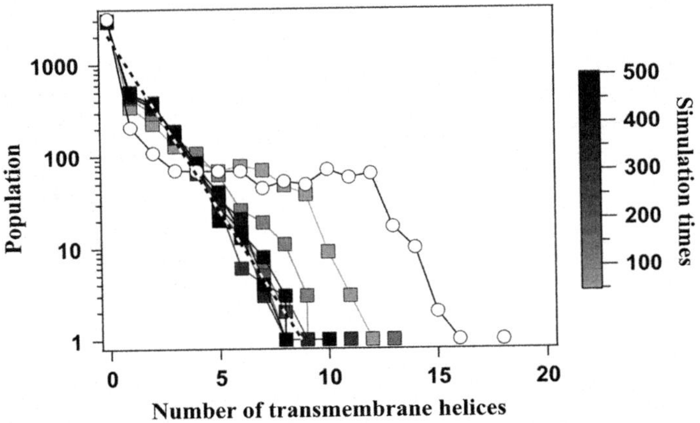

Fig. 10.6 Change in the distribution of the number of transmembrane helices of membrane proteins when the mutation simulation was performed with the nucleotide composition fixed to that of the actual genome. The prominent shoulders gradually disappeared, and the distribution became a single exponential function. From Sawada and Mitaku (2011)

to investigate how the distribution of membrane proteins changes with the number of transmembrane helices (Sawada and Mitaku 2011). In the actual *E. coli* genome, the protein distribution clearly showed a shoulder at approximately 12 transmembrane helices. This shoulder at approximately 12 helices was also observed in other prokaryotes. However, when simulations were performed while ignoring the function, the shoulders gradually disappeared and eventually converged to a single exponential distribution (Sawada and Mitaku 2011).

As mentioned earlier, a single exponential distribution is generally a feature of random processes. Therefore, it is quite reasonable that a single exponential distribution was obtained by a random mutation simulation ignoring function (Sawada and Mitaku 2011). In contrast, the presence of prominent shoulders in the number distribution of transmembrane helices in real genomes suggests that membrane proteins near these prominent shoulders have important functions. Nonetheless, it is very interesting to note that the proportion of total membrane proteins is constant, at approximately 1/4 (Fig. 10.6).

10.2 Conversion of Membrane Proteins and Soluble Proteins in the Process of Evolution

As the simulation results show, the proportion of membrane proteins is determined by the probability of mutation (nucleotide composition). It is conceivable that many mutations have occurred during the course of evolution, resulting in a conversion between soluble and membrane proteins. Then, this conversion can be described in the form of a reaction equation (Fig. 10.7). Let $k_{s \to m}$ be the reaction rate at which a mutation converts a soluble protein to a membrane protein and $k_{m \to s}$ be the reaction

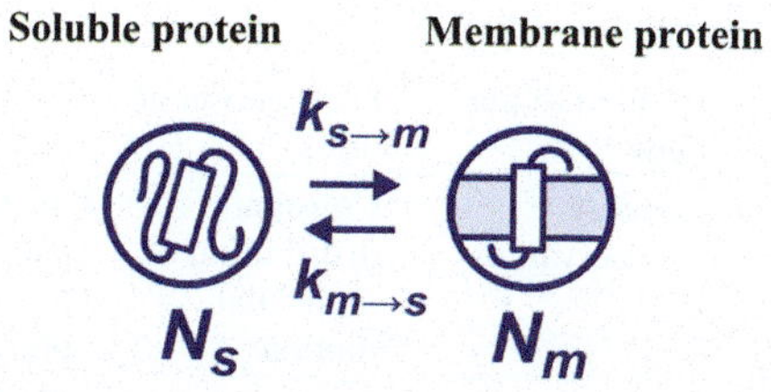

Fig. 10.7 Reaction equation showing that the distribution of membrane proteins is constant at equilibrium. The reaction rate $k_{s \to m}$ for the conversion reaction from soluble protein to membrane 99protein and the reaction rate $k_{m \to s}$ for the reverse reaction are both reaction rates in the evolutionary process. From Mitaku (2013)

rate at which a mutation converts a membrane protein to a soluble protein. Then, when the reaction of this conversion is in equilibrium, the relationship between the number of membrane proteins N_m and the number of soluble proteins N_s is as follows:

$$k_{s \to m} N_s = k_{m \to s} N_m \tag{10.1}$$

That is, in equilibrium, the number of membrane proteins converted to soluble proteins and the number of soluble proteins converted to membrane proteins are the same. Thus, as shown in Eq. (10.2), the fraction of membrane proteins is determined solely by the ratio of reaction rates.

$$R = \frac{N_m}{N_s + N_m} = \frac{k_{s \to m}}{k_{s \to m} + k_{m \to s}} \tag{10.2}$$

Intracellular processes involve a series of molecular devices, such as translocons (Hessa et al. 2005), that embed proteins in the membrane. Furthermore, the same devices are used to make membrane proteins in a variety of organisms. The fraction of membrane proteins that is determined by the conversion between membrane proteins and soluble proteins is then determined solely by the ratio of reaction rates, as in Eq. (10.2). Here, the rate of mutation varies greatly depending on the environment of the organism, but if the ratio of reaction rates is the same, the ratio of membrane proteins remains the same. This mechanism is thought to keep the ratio of membrane proteins constant among organisms (Mitaku 2013).

10.3 Conservation Laws for Nucleotide Composition in Genome Sequences

As shown in the simulations in the first half of this chapter, the fraction of membrane proteins is completely dependent on the nucleotide composition at each codon position (Sawada and Mitaku 2012). When the nucleotide composition is fixed to the actual value in the simulation, the fraction of membrane proteins matches the

Table 10.1 Mechanisms of stability in matter and organisms (2)

System	Elementary process	Conservation law	Distribution of units	Parameters for phase diagram	State
Matter (e.g. gas)	Random motion of molecules	Energy conservation	Constant distribution of molecular motion	Intensive parameters (e.g. temperature and pressure)	Three states of matter
Organism	Random mutations to genome sequences	Conservation of nucleotide compositions	Constant distribution of different types of proteins	Nucleotide compositions for first and second letters of codon	Ordered state of life

From the analysis of the genome sequence, the distribution by protein type is constant, and the parameter for the phase diagram is the nucleotide composition by codon letter position. Furthermore, conservation of nucleotide composition was established

actual value. In other words, nucleotide composition is conserved in real organisms, resulting in a constant protein distribution. Knowing this fact, we can complete the similarity table of substances and organisms, as shown in Table 10.1.

First, elementary processes in matter are random molecular motions, while elementary processes in biological genomes are random mutations. In both systems, elementary processes are random. Second, if the overall random motion is in equilibrium, energy conservation is established. On the other hand, in living organisms, nucleotide composition is considered to be conserved despite random mutations. Finally, if energy conservation holds in matter, the distribution of motion is constant (e.g., Maxwell distribution). On the other hand, if the nucleotide composition of the biological genome is in equilibrium, the fraction of membrane proteins is constant. In other words, if random processes are in equilibrium in both material and biological systems, the distribution of random processes is constant. Thus, the similarity between material and biological systems is valid in most respects.

The remaining entries in Table 10.1 concern the phase diagram of life. Deeply related to this problem is the low-entropy state of biological systems. Organisms as a whole are considered to have very low entropy (Schroedinger 1944; Monod 1971). However, the low-entropy state of organisms has not been discussed quantitatively. If we can draw a phase diagram of life from the genome sequence, we may be able to determine the cause of low entropy from the region where organisms can exist. Thus, in the next chapter, we will discuss the phase diagram of life and the low entropy of organisms.

References

Hessa T, Kim H, Bihlmaier K, Lundin C et al (2005) Recognition of transmembrane helices by the endoplasmic reticulum translocon. Nature 433:377–381

Mitaku S (ed) (2013) Computational science and engineering, Genome computing; toward genome reality beyond bioinformatics, vol 7. Kyoritsu-Shuppan Co., Tokyo (in Japanese)

Mitaku S (2015) A modern approach to biological science. Kyoritsu-Shuppan Co., Tokyo (in Japanese)

Monod J (1971) El hansard et la necessite. Alfred A. Knopf, Inc., Paris
Sawada R, Mitaku S (2011) Number distribution of transmembrane helices in prokaryote genomes. In: Computational biology and applied bioinformatics, pp 279–286
Sawada R, Mitaku S (2012) Biological meaning of DNA compositional biases evaluated by ratio of membrane proteins. J Biochem 151:189–196
Sawada R, Ke R, Tsuji T, Sonoyama M, Mitaku S (2007) Ratio of membrane proteins in total proteomes of prokaryota. Biophysics 3:37–45
Schroedinger E (1944) What is life? The physical aspect of the living cell. Cambridge University Press, Cambridge

Chapter 11
Habitable Zone in Nucleotide Composition Space: Phase Diagram of Life

Keywords Phase diagram of life · Habitable zone · Nucleotide composition space · Intensive parameter · Extreme environmental microorganisms

We plotted the whole-genome sequences of many organisms in nucleotide composition space. As a result, all genome sequences were distributed in a very narrow region in the four-dimensional nucleotide composition space of each of the first and second letters of the codon. In other words, the eight-dimensional compositional space of the first and second letters of the codon represents the habitable zone, the area where organisms can survive. A major characteristic of the habitable zone is that it deviates significantly from a completely random composition. In such cases, the number of possible DNA sequences is fiercely small, which naturally explains why the organism is in an extremely low-entropy state. Furthermore, DNA sequences that would make it impossible to design an organism are eliminated in advance, and the mutations that occur in an organism are naturally neutral. Furthermore, physical considerations have shown that the region of nucleotide composition called the habitable zone is the phase diagram of life.

11.1 The Distribution of Genomes in Nucleotide Composition Space Is Highly Biased

From the discussion in the previous chapters, we have seen that the nucleotide composition of each letter of the codon was an important parameter in discussing the macroscopic properties of the organism. Since a codon corresponding to an amino acid is composed of three nucleotides and there are four different nucleotides at each codon position, it is the nucleotide composition of the 12 letters that characterizes the genome sequence. However, each letter position of a codon can be considered to have a different physical meaning, at least statistically. We first plotted the genomes of many organisms in four-dimensional space corresponding to each letter position of the codon. The results are shown in Figs. 11.1, 11.2, and 11.3 (Mitaku and Sawada 2018).

S. Mitaku, R. Sawada, *Evolution Seen from the Phase Diagram of Life*, Evolutionary Studies, https://doi.org/10.1007/978-981-97-0060-8_11

To represent a four-dimensional space in a two-dimensional graph, six graphs are needed. Thus, each figure shows six graphs for the position of each letter. First, if we look at the plot of the compositional space for the third letter (Fig. 11.3), each nucleotide composition is characterized by a large spread in the range of approximately 0–0.5. Since a completely random composition is 0.25, the spread over the range of 0–0.5 means that almost any composition is possible for the third letter. In this sense, there is no bias in the base composition of the third letter of the codon.

On the other hand, for the first and second letters of the codon, all genomic sequences were in a very narrow region compared to the third letter (Figs. 11.1 and 11.2). Furthermore, when the plots of the first and second letters in compositional space were compared, their positions were very different. For example, the composition of thymine in the second letter was greater than 0.25, whereas the composition of thymine in the first letter was less than 0.25. In contrast, the composition of guanine in the second letter was less than 0.25, while the composition of guanine in the first letter was greater than 0.25. These trends are common to all 2664 organisms, strongly suggesting that organisms cannot survive outside of these regions. Notably, both eukaryotes (blue) and prokaryotes (gray and black) are found in the same region. In addition, both normal environmental microorganisms (black) and extreme environmental microorganisms (gray) are present in the same region,

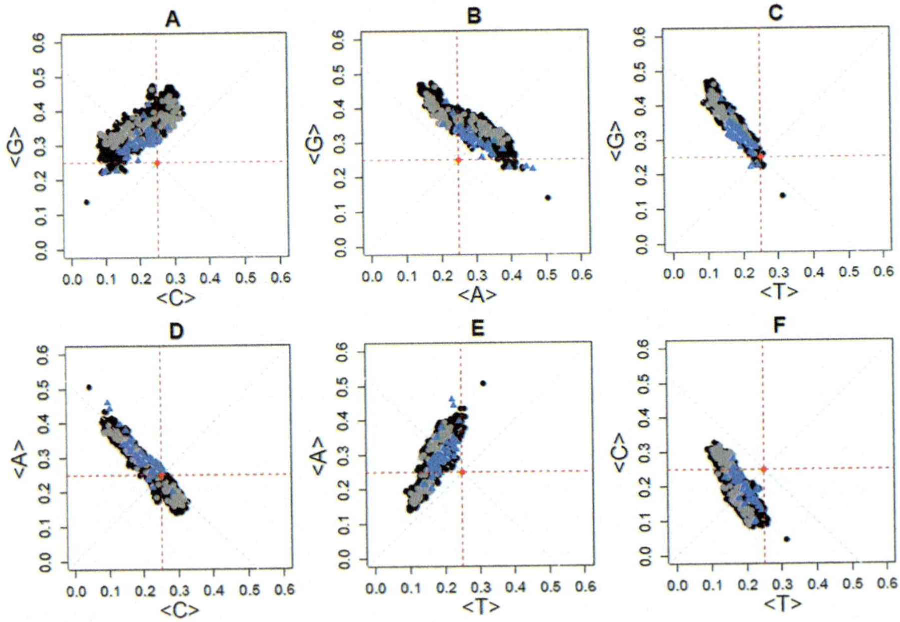

Fig. 11.1 Six two-dimensional scatter plots showing the four-dimensional space of nucleotide composition of the first letter of the codoncodon: (**a**), <C> vs <G>; **b**, <A> vs <G>; **c**, <T> vs <G>: **d**, <C> vs <A>; **e**, <T> vs <A> and **f**, <T> vs <C>. A total of 173 thermophilic bacteria (gray), 2378 other prokaryotes (black), and 113 eukaryotes (blue) are plotted. All 2664 species are plotted in a narrow region, strongly suggesting a phase diagram of life. From Mitaku and Sawada (2018)

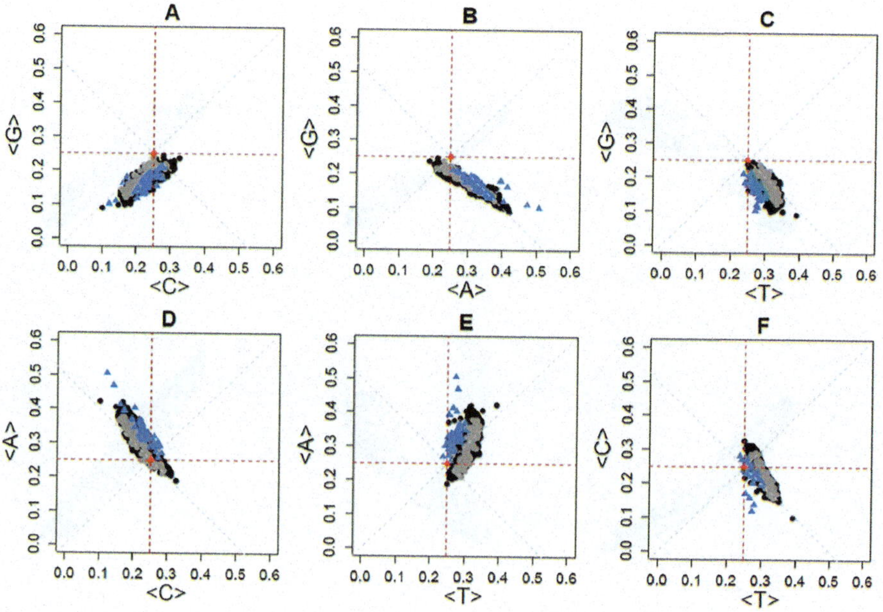

Fig. 11.2 Six two-dimensional scatter plots showing the four-dimensional space of nucleotide composition of the second letter of the codon: **a**, <C> vs <G>; **b**, <A> vs <G>; **c**, <T> vs <G>; **d**, <C> vs <A>; **e**, <T> vs <A> and **f**, <T> vs <C>. The plotted species are the same as in Fig. 11.1. Again, all 2664 species are plotted in a small region, strongly suggesting a phase diagram of life. From Mitaku and Sawada (2018)

indicating that the survival environment is independent of their position in compositional space.

The fact that the nucleotide compositions of the first and second letters are plotted in very different regions can be easily understood by looking at the arrangement of amino acids in the codon table (Fig. 1.3). The group of amino acids specified by the first letter of the codon is statistically related to the flexibility of the amino acid segment. On the other hand, the group of amino acids designated by the second letter is related to the hydrophobicity of the amino acid segment. In other words, the nucleotide composition of the first and second letters may determine the average flexibility and hydrophobicity of all proteins by the genome, respectively. Thus, it is well understood that the nucleotide compositions of the first and second letters are in different regions when viewed through the physical properties of the amino acids.

Thus, in the sections and chapters that follow, we will examine in detail both the physical and biological significance of the narrow region in nucleotide composition space. Physically, the fact that all organisms are plotted in a narrow region of compositional space may mean that we have obtained a phase diagram of life. We named this region the habitable zone, and the relationship between the location of the habitable zone in compositional space and the distribution of proteins is a very interesting question (Mitaku and Sawada 2018).

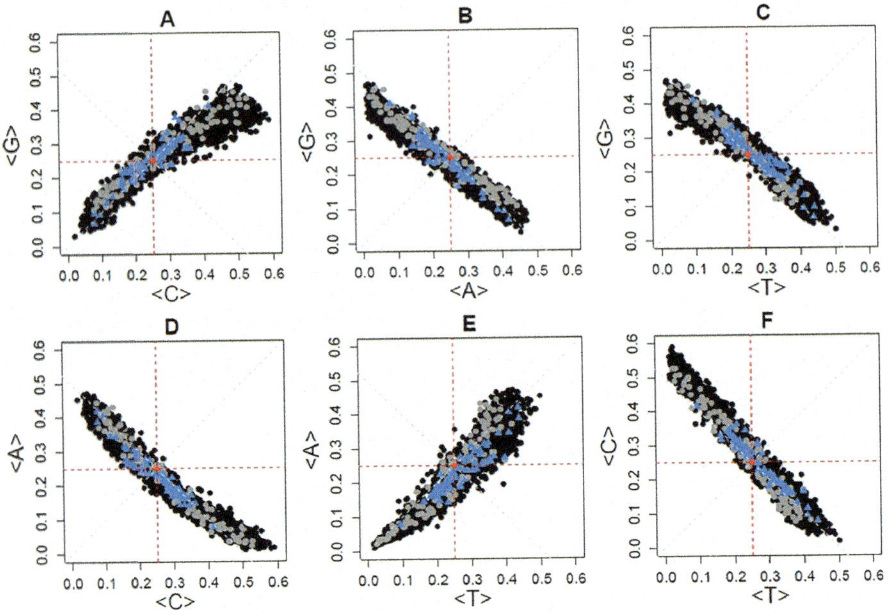

Fig. 11.3 Six two-dimensional scatter plots showing the four-dimensional space of nucleotide composition of the third letter of the codon: **a**, <C> vs <G>; **b**, <A> vs <G>; **c**, <T> vs <G>; **d**, <C> vs <A>; **e**, <T> vs <A> and **f**, <T> vs <C>. Plotted species are the same as in Fig. 11.1. For the third letter of the codon, the plots are so widely distributed that this cannot be assumed to be part of the phase diagram of life. From Mitaku and Sawada (2018)

11.2 The Physical Meaning of the Highly Biased Nucleotide Composition in Genomes

In general, the number of possible sequences depends on how far the composition of each nucleotide deviates from a completely random composition (0.25). If the composition of one nucleotide is 1 and the composition of all other nucleotides is 0, there is only one possible sequence. On the other hand, if the composition of each nucleotide is 0.25, then the number of possible sequences is maximized. In other words, as the bias increases from a completely random composition, the number of possible sequences decreases correspondingly. The degree of decrease in the number of possible sequences depends on the distance from the perfectly random composition.

Therefore, we calculated the distance between the actual genomic sequence composition and the completely random composition for the first and second letters of the codon. In Eqs. (11.1) and (11.2), d_1 and d_2 are the distances at the first and second letter, respectively. Here, x_1 and x_2 are the nucleotide composition values at the first and second letter, respectively.

$$d_1 = \sqrt{\sum_{i=A,T,G,C} \left(x_1(i) - 0.25\right)^2} \tag{11.1}$$

$$d_2 = \sqrt{\sum_{i=A,T,G,C} \left(x_2(i) - 0.25\right)^2} \tag{11.2}$$

Furthermore, the total distance D_{12} for d_1 and d_2 combined can be calculated as

$$D_{12} = \sqrt{d_1^2 + d_2^2} \tag{11.3}$$

First, we plotted the nucleotide composition of the genome on a scatter plot of d_1 vs. d_2 (Fig. 11.4). Plotted are 2551 prokaryotes (circles) and 113 eukaryotes (triangles). Figure 11.4 is a two-dimensional version of the eight-dimensional habitable zone calculated by Eqs. (11.1) and (11.2). The graph clearly shows that all genomes are located far from the origin (a completely random composition).

In Fig. 11.4, prokaryotes are shown in grayscale according to their GC content. Each organism systematically and gradually shifts its position according to its GC content. In other words, each species has a preferred nucleotide composition according to its GC content. In general, organisms that are evolutionarily close to each

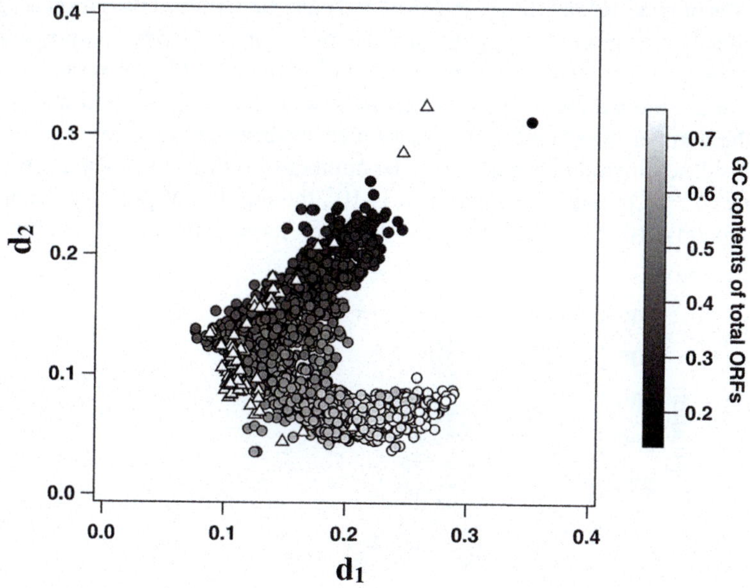

Fig. 11.4 Euclidean distance between prokaryotic genome composition and completely random composition (scatterplot of d_1 vs. d_2). Here, d_1 and d_2 are the distances between the first and second letters of the codon calculated by Eqs. (11.1) and (11.2), respectively. The composition of all biological genomes deviates significantly from a completely random composition. The circles are shown in grayscale according to GC content of prokaryotes, indicating that each species is phylogenetically displaced in compositional space. Triangles represent eukaryotes

other have similar DNA sequences and, naturally, close nucleotide compositions, which naturally leads to close GC content. This suggests that the evolutionary process is a random walk in the habitable zone.

Moreover, Fig. 11.4 shows that all genome sequences deviate significantly from the origin (a completely random composition). The genome with a GC content of approximately 0.5 is plotted closest to the origin, but even then, it deviates significantly from a completely random composition. The distance between the origin and the habitable zone can be calculated using Eq. (11.3), and the resulting histogram is shown in Fig. 11.5.

This graph clearly shows a rather sharp peak at a distance of approximately 0.2. Here, the black bars represent prokaryotes, and the blue bars represent eukaryotes. This graph means that all actual genomes are at a distance of approximately 0.2 from a completely random composition (Mitaku and Sawada 2018). As mentioned earlier, the number of possible sequences decreases with distance from a completely random composition, and the extent of the decrease in the number of possible sequences can be roughly assessed by scaling the distance by its standard deviation.

Assuming that there are N random processes selecting one of the four nucleotides, the standard deviation is as follows:

$$\delta = \sqrt{3/16N} \tag{11.4}$$

where N is at least 10^6, as shown in Fig. 9.5. Then, the standard deviation is approximately 4×10^{-4} or more. Scaling the distance of 0.2 in nucleotide composition space by the standard deviation corresponds to approximately 500σ or more. As is well known, at 2σ, the number of possible sequences decreases to approximately 5%. At 500σ, the number of possible genomic sequences drops drastically.

In this case, the reduction factor in the number of possible sequences is approximately $e^{-125,000}$. On the other hand, if N is 10^6, the number of possible sequences is approximately 10^{600000}. Thus, despite the very large reduction, there are still a

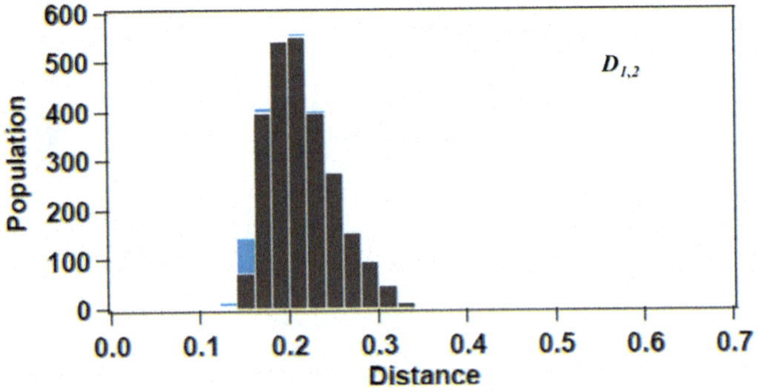

Fig. 11.5 Histogram of the Euclidean distance (Eq. 11.3) between actual genome composition and complete random composition in nucleotide composition space. All biological genome compositions are distributed at a distance of approximately 0.2 from the completely random composition. Blue bars represent eukaryotes. From Mitaku and Sawada (2018)

sufficient number of sequences remaining! The excluded sequences are nonviable sequences, while the remaining sequences in the habitable zone should be mostly viable sequences. In other words, while reducing the number of unwanted sequences, there is still plenty of potential for future evolution.

11.3 Nucleotide Composition Space Meets the Requirements of a Physical Phase Diagram

The properties of the eight-dimensional nucleotide composition space (Figs. 11.1 and 11.2) or the two-dimensional scatterplot of distances in composition space (Fig. 11.4) strongly suggest that they correspond to the phase diagram of life. However, for these to be physical phase diagrams, at least two conditions must be met. One is that the phase diagram of life must be a sufficiently narrow domain that includes all organisms. The other is that the parameters of the phase diagram of life must be intensive parameters. Below we would like to discuss these conditions.

- Since nucleotide composition is expected to vary among different species, genera, families, etc., we plotted many genomes in the eight-dimensional base composition space of the first and second letters of the codon (Figs. 11.1 and 11.2). There, all genomes were in a narrow region. We also plotted a two-dimensional scatter plot of distances in the compositional space of the genome sequence (Fig. 11.4), and all the organisms we examined were located well outside of a completely random composition. It is important to note here that both prokaryotes and eukaryotes, which differ widely in genome sequence, are in the same narrow region. Furthermore, extreme and normal environmental microorganisms are also plotted in the same narrow region. Although the organisms examined in this study are few in number compared to all organisms on Earth, the fact that very different types of organisms are all in the narrow region without exception fulfills the first condition of the phase diagram of life (Mitaku and Sawada 2018).
- Let us consider the physical point of view of the phase diagram of life. According to Fermi's thermodynamics, to draw a phase diagram for the state of matter, one must use intensive parameters (Fermi 1936). For example, temperature and pressure are intensive parameters, while the volume and number of molecules are extensive parameters. To draw a phase diagram, at least one parameter must be an intensive parameter. In the case of a genome sequence, the DNA sequence itself and the set of genes are extensive parameters, as is the volume in the case of matter. Thus, the set of genes alone is not enough to draw a phase diagram of an organism. In contrast, nucleotide composition is an intensive parameter, and from a physical point of view, nucleotide composition is a reasonable parameter for a phase diagram.

In this chapter, we have shown that the conservation law of nucleotide composition holds for genome sequences and that the phase diagram of life can also be depicted by nucleotide composition. However, the relationship between the conservation law of nucleotide composition and the constant distribution of proteins is not yet clear. Therefore, in the next chapter, we will consider the biological and physical relationship between the conservation laws of nucleotide composition (or the phase diagram of life) and the distribution of proteins.

References

Fermi E (1936) Thermodynamics. Dover publications, Inc., New York

Mitaku S, Sawada R (2018) Biological meaning of "habitable zone" in nucleotide composition space. Biophys Physicobiol 15:75–85

Chapter 12
Relationship Between the Phase Diagram of Life and Protein Distribution

Keywords Phase diagram of life · Mode of fluctuations · Percentage of membrane proteins · Molecular recognition · Intracellular factors

A significant bias in the nucleotide composition of the first and second letters of the codon is a major characteristic of living organisms. There is a deep relationship between the bias in nucleotide composition and the constant distribution of proteins described in Chaps. 9 and 10. Fluctuations in the DNA sequence due to random mutations determine the distribution of proteins, and the process is as follows. When the nucleotide composition deviates significantly from a completely random composition, a unique mode of fluctuation occurs in the DNA sequence. For example, a bias in the nucleotide composition of the second letter of a codon will result in a large hydrophobic and amphiphilic fluctuation in the amino acid sequence due to the bias in the physical properties of the amino acids in the codon table. When the mode of this fluctuation reaches equilibrium, the proportion of membrane proteins is constant (approximately 1/4). The first letter of the codon, on the other hand, relates to the structural freedom of the amino acid segment, which generates the molecular recognition unit of the protein. In other words, if nucleotide composition is maintained within the phase diagram of life, membrane proteins and molecular recognition units will occur in approximately constant proportions. At the end of this chapter, we also discuss the intracellular factors that maintain a biased nucleotide composition.

12.1 Relationship Between Fluctuations in DNA Sequences and Physical Properties of Amino Acid Sequences

The whole-genome sequence is more than just a set of genes. In Chap. 11, by analyzing the entire coding region of the genome sequence, we showed that the nucleotide composition at each codon position deviates significantly from a completely random composition. In other words, organisms have always chosen compositions that deviate greatly from completely random compositions. Furthermore, in Chap.

S. Mitaku, R. Sawada, *Evolution Seen from the Phase Diagram of Life*, Evolutionary Studies, https://doi.org/10.1007/978-981-97-0060-8_12

10, we showed that only when fixed to the actual nucleotide composition did the percentage of membrane proteins simulated by the random mutation remain constant at approximately one-fourth. This suggests that a significant bias in nucleotide composition can result in a constant protein distribution, which is very advantageous for life.

Figure 12.1 shows a model flowchart of the conversion from fluctuations in the nucleotide sequence of DNA to fluctuations in the physical properties of amino acid segments. In these graphs, window averages with a width of approximately 10 residues are plotted. Subscripts 1 through 4 are adenine (A), thymine (T), guanine (G), or cytosine (C). To simplify the logic, assume that the composition of nucleotides 3 and 4 is small and constant. Then, the compositions of the four nucleotides add up to one, so the variation in the probability of occurrence of nucleotides 1 and 2 will be large, and they will change in opposite directions. That is the left side of Fig. 12.1. We can then consider how the physical properties of the amino acid segment change as the composition of nucleotides 1 and 2 change locally.

The graph on the right side of Fig. 12.1 shows the trend of the physical properties of the amino acid sequence when the nucleotide sequence on the left side is translated. Since the trend of the physical properties of the amino acid sequence is completely different between the first and second letters of the codon, the right side shows two separate graphs, one for the first letter and one for the second letter. The

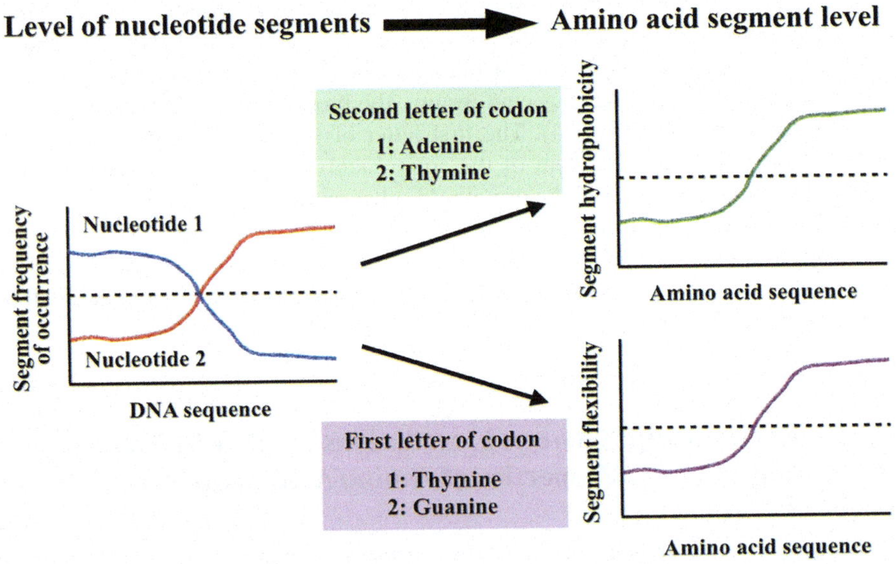

Fig. 12.1 Flowchart showing the conversion from the distribution of nucleotide composition to the distribution of physical properties of amino acid sequences by window averaging. Large changes in nucleotide composition at the first and second letters of the codon correlate well with the distribution of physical properties of the amino acid sequence. The dominant physical properties of the amino acid sequence are quite different for the changes in nucleotide composition at the first and second letters of the codon

amino acid sequence itself cannot be determined from the nucleotide composition information alone, but since there is a large bias in the physical properties of amino acids in the codon table, it is possible to show changes in the physical properties of amino acid segments from changes in nucleotide composition.

First, for the second letter of the codon, if nucleotide 1 is adenine and nucleotide 2 is thymine, the physical properties of the amino acid segment will change from low to high hydrophobicity. In contrast, in the case of the first letter of the codon, if nucleotide 1 is thymine and nucleotide 2 is guanine, the physical properties of the segment would change from less to more flexible.

As more mutations are introduced throughout the genome sequence, different modes of nucleotide sequence fluctuations occur. Eventually, when these various modes reach equilibrium, their distribution becomes constant. The fluctuations in the frequency of nucleotide occurrences are then converted into fluctuations in the physical properties of the amino acid sequence by the mechanism shown in Fig. 12.1. In equilibrium, the mode of fluctuation of the physical properties of the amino acid sequence becomes constant. As a result, the protein distribution also becomes constant. We illustrate this process in the following two sections.

12.2 Fluctuations in Nucleotide Composition at the Second Letter of the Codon Keep the Proportion of Membrane Proteins Constant

As Anderson (1972) stated, in a system composed of many elements, the properties of the individual elements are quite different from the properties of the system as a whole. Similarly, the properties of the genome as a whole should be quite different from the properties and functions of individual genes. When many mutations are introduced into the genome sequence, fluctuations occur in various modes with respect to the frequency of occurrence of each nucleotide. In equilibrium, however, the frequency of occurrence of the fluctuating modes will be constant. This is what causes the protein distribution to be constant. As an example, we will first discuss the mode of nucleotide fluctuation in the second letter of the codon (Mitaku and Sawada 2018).

Figure 12.2a shows some of the nucleotide compositions at the second letter of the codon. Here, the thymine composition is greater than 0.25. The composition of adenine is also generally greater than 0.25. The composition of guanine is less than 0.25. Next, Fig. 12.2b shows the codon table, with hydrophobic amino acids in blue and amphiphilic amino acids in red. Therefore, combining Fig. 12.2a and b, we see that, as an average for all proteins, hydrophobic and amphiphilic amino acids are more abundant than average. On the other hand, there should be fewer secondary structure-breaking amino acids (proline, glycine, and other small amino acids), and the secondary structure should be longer. Thus, in equilibrium, there will be more segments with the mode of relatively long hydrophobic segments flanked by amphiphilic segments. Such a mode of fluctuation is shown in Fig. 12.2c, which actually

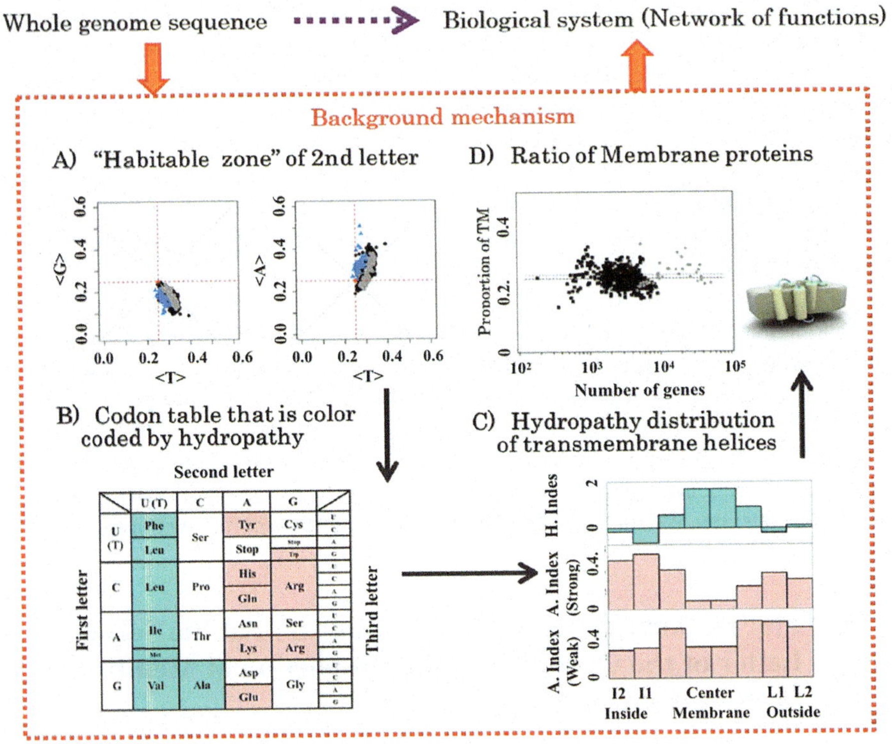

Fig. 12.2 Flowchart showing that membrane proteins are determined by fluctuations in the nucleotide composition of the second letter of the codon. (**a**) Distribution of the nucleotide composition of the second letter. (**b**) Arrangement of hydrophobic (blue) and amphiphilic (red) amino acids in the codon table. (**c**) Relationship between hydrophobic and amphiphilic clusters forming transmembrane helices (SOSUI system algorithm). (**d**) Diagram showing the constant proportion of membrane proteins in the genome sequence. The function of each protein is considered to be determined after this flow. From Mitaku and Sawada (2018)

corresponds to the transmembrane helix of Fig. 9.1. In other words, Fig. 12.2c is the algorithm of SOSUI (Mitaku & Hirokawa 1999; Mitaku et al. 2002; Tsuji & Mitaku 2004), a membrane protein prediction system with more than 95% accuracy (Hirokawa et al. 1998).

Figure 12.2d shows that the percentage of membrane proteins in the total amino acid sequence from the genome is constant (Fig. 9.3). Thus, the entire flowchart in Fig. 12.2 shows that the percentage of membrane proteins is constant due to the highly skewed nucleotide composition. This shows that the distribution of proteins can be kept constant simply by controlling the nucleotide composition. Then, by simply adding a few amino acids in the active site of function, we can create a protein with function. Thus, it follows that by simply controlling the nucleotide composition, an organism can eliminate a large number of unwanted sequences and efficiently obtain a group of proteins.

12.3 Fluctuations in Nucleotide Composition at the First Letter of the Codon Form the Molecular Recognition Site

In the distribution of nucleotide compositions at the first letter of the codon, the composition of guanine is greater than 0.25 and that of thymine is less than 0.25. Amino acids corresponding to the first letter guanine include glycine, alanine, and valine, which are all small side chains. In addition, aspartic acid and asparagine, which correspond to the first-letter guanine, are the smallest of the hydrophilic amino acids. As a result, a larger composition of the first-letter guanine results in a higher average degree of freedom of the amino acid segment, making it mechanically softer (Mitaku and Sawada 2018).

In contrast, if the first letter of the codon is thymine or cytosine, many of the corresponding amino acids, such as phenylalanine, tryptophan, tyrosine, and histidine, have cyclic side chains with large volumes, making the segment stiffer. In addition, cysteine, which forms S–S bonds, and proline, an imino acid, greatly reduce the degrees of freedom of the protein structure (Branden and Tooze 1998). Many of the proteins that make up living organisms are known to be structurally soft, which is consistent with the large composition of guanine and small composition of thymine and cytosine in the first letter of the codon. Furthermore, the fluctuations in the sequence should result in the formation of many flexible segments.

Therefore, let us consider what segmental flexibility means for the genome as a whole. One study on this topic is the relationship between the molecular recognition sites of allergens and amino acid fluctuations. An allergen is a protein that is molecularly recognized by an immune protein and activates an allergic response. Allergenic amino acid segments are thought to be present in allergens, but the same amino acid segments are absent in nonallergenic proteins. To statistically analyze these phenomena, we used a database of allergenic and nonallergenic amino acid sequences. From these databases, we listed amino acid sequences of 3–8 residues that are present only in allergens. We then characterized the amino acid sequences surrounding these segments (Asakawa et al. 2010).

When examining the probability of occurrence of amino acids around the fragments found only in allergens, two types of amino acids were found that show contrasting distributions. As shown in the codon table in Fig. 12.3b, one was the group of amino acids shown in yellow-green (glycine, alanine, aspartate, glutamic acid, and lysine), and the other was the group shown in light blue (phenylalanine, tyrosine, tryptophan, cysteine, histidine, and proline). Amino acids in the yellow-green group appeared more abundant in the center of the wavelet and decreased on both sides (Fig. 12.3c). On the other hand, the amino acids in the light blue group showed the opposite distribution, with lower frequencies in the center and higher frequencies on both sides. The remaining amino acids were simply white noise. In other words, the allergen epitope (allergen binding site) showed just such a distribution of amino acids (Fig. 12.3d) (Asakawa et al. 2010).

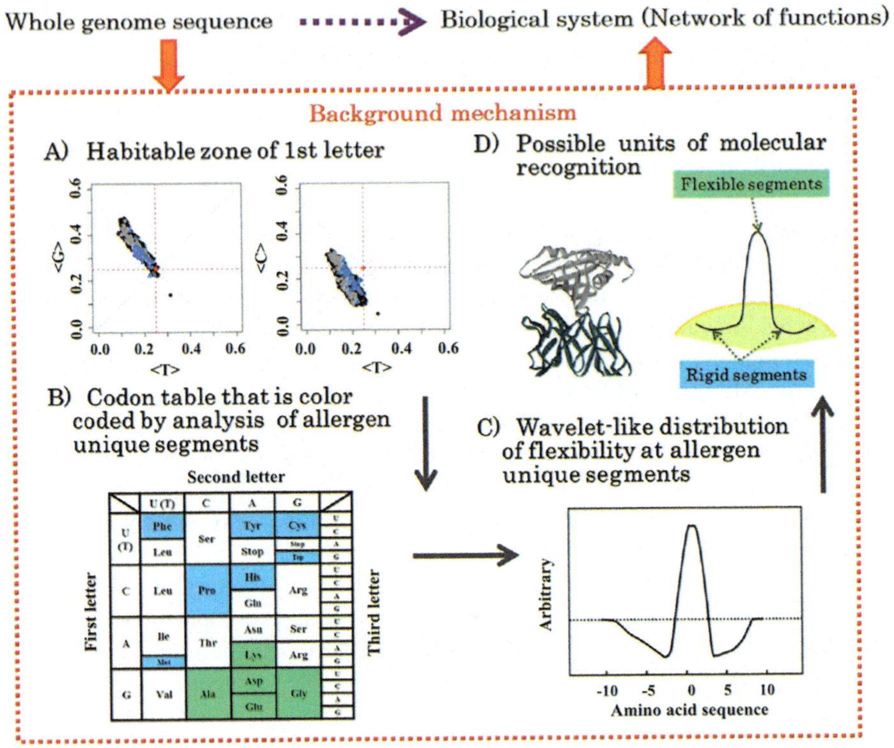

Fig. 12.3 Flowchart suggesting that the molecular recognition unit is formed by fluctuations in the nucleotide composition of the first letter of the codon. (**a**) Distribution of nucleotide composition of the first letter. (**b**) The arrangement of flexible (yellow-green) and rigid (light blue) amino acids in the codon table. (**c**) Relationship between clusters of flexible amino acids and clusters of rigid amino acids forming the molecular recognition unit. (**d**) Model of the formation of the molecular recognition unit and the three-dimensional structure of the pollen allergen. The function of each protein is thought to be determined after this process. From Mitaku and Sawada (2018)

Let us consider this problem back to the nucleotide composition again. As shown in Fig. 12.3a, the composition of guanine is larger in the first letter of the codon. As a result, there are more yellow-green amino acids in the codon table. In other words, these segments are softer. On the other hand, the frequency of thymine and cytosine in the first letter is smaller. This would result in fewer light blue amino acids at the codon. In other words, the number of amino acids that make the segment hard is relatively small. Then, as shown in Fig. 12.3c, the fluctuation caused by the mutation generates many flexible amino acid fragments. Thus, a large bias in the nucleotide composition of the first letter of the codon alone will cause many wavelet-like flexible fluctuations.

If the fist is held tightly with only the index finger out, the index finger moves flexibly over the firm fist. This forms a unit that gently grips objects. The molecular recognition unit of a protein is similar to this shape. The active site is formed by combining several such units. Because the composition of the first letter of the codon is highly skewed, random mutations automatically result in the formation of many molecular recognition units. The molecular recognition sites should then form spontaneously (Fig. 3.8).

The frequency of occurrence of molecular recognition units can be roughly evaluated from the amino acid sequence. A trial calculation shows that, on average, approximately one molecular recognition unit appears per 12 residues. This means that for a single protein, a significant number of molecular recognition units can be created. The relatively hydrophobic molecular recognition units are then positioned inside the protein, while the relatively hydrophilic units are positioned outside the protein. The former would form the three-dimensional structure of the protein, while the latter would bind to other molecules. The amino acid sequence near the epitope of the allergen shows such a tendency.

Although this study is not fully advanced, we believe that the bias in the nucleotide composition of the first letter of the codon forms the molecular recognition unit.

12.4 What Determines the Bias in Nucleotide Composition?

The flowcharts in Figs. 12.2 and 12.3 suggest that the constant appearance of the percentage of membrane proteins and molecular recognition units is due to random mutations that preserve the bias in the nucleotide composition of the genome sequence. In other words, nucleotide compositional bias is a prerequisite for the formation of biological systems. The final fundamental question in this chapter is what factors maintain the large bias in nucleotide composition? This issue has already been discussed in Chap. 8 (Fig. 8.1). However, we would like to discuss this issue again because it is crucial to the discussion of biological evolution.

During evolution, various mutations are introduced into the genome sequence. However, various repair enzymes work on the mutations, and approximately 999 out of 1000 mutations are correctly repaired. However, approximately 1 out of 1000 mutations is repaired incorrectly, creating a new genome. Genomic sequences with some mutations will be translated and form a new biological system. If the biological system survives, the genome sequence will be passed on to the next generation. As shown in Fig. 8.1, there are two processes that can potentially control the nucleotide composition of the next generation's genomic sequence: repair system bias and natural selection against mutations.

Repair enzymes are very good and repair most mutations correctly. However, mistakes occasionally occur during the repair process. It is not known whether the nucleotide composition due to these mistakes is completely random. However, if there is a bias in the frequency of nucleotide occurrences due to repair errors, then the nucleotide composition of the entire genome will naturally be biased. This is the cause of the bias in nucleotide composition and, as discussed in the first half of this chapter, has great potential to control protein distribution (Mitaku and Sawada 2018).

Another possible source of bias in nucleotide composition is the process of natural selection in evolution. This is the effect of external environmental pressures. It has been established, however, that mutations in individual genes are generally neutral (Kimura 1968; King and Jukes 1969; Saitou 2009). The fact that gene mutations are neutral is equivalent to the prior exclusion of a vast number of DNA sequences that cannot form an entire organism. In other words, it is reasonable to assume that the bias in nucleotide sequences is caused by a bias in the repair system, resulting in the natural elimination of many useless sequences, thereby making natural selection neutral.

Despite their complexity, organisms are very stable systems that allow them to undergo great changes in evolution. We have never realized that intracellular factors (perhaps biases in repair systems) are responsible for the stability and evolution of organisms. However, we believe that this fact is the cause of the great mystery of the organism. In Part III, we discussed in detail the nucleotide composition from the genome sequence and revealed the phase diagram of life. In Part IV, we will discuss how this phase diagram of life and intracellular factors for bias in nucleotide composition can explain the mysteries of evolution.

References

Anderson P (1972) More is different. Science 177:393–396

Asakawa N, Sakiyama N, Teshima R, Mitaku S (2010) Characteristic amino acid distribution around segments unique to allergens. J Biochem 147:127–133

Branden C, Tooze J (1998) Introduction to protein structure, 2nd edn. Garland Science, New York, NY

Hirokawa T, Seah B-C, Mitaku S (1998) SOSUI: classification and secondary structure prediction system for membrane proteins. Bioinformatics 14:378

Kimura M (1968) Evolutionary rate at the molecular level. Nature 217:624–626

King JL, Jukes TH (1969) Non-Darwinian evolution: Most evolutionary change in proteins may be due to neutral mutations and genetic drift. Science 164:788–798

Mitaku S, Hirokawa T (1999) Physicochemical factors for discriminating between soluble and membrane proteins: hydrophobicity of helical segments and protein length. Protein Eng 12:953–957

Mitaku S, Sawada R (2018) Biological meaning of " habitable zone" in nucleotide composition space. Biophys Physicobiol 15:75–85

Mitaku S, Hirokawa T, Tsuji T (2002) Amphiphilicity index of polar amino acids as an aid in the characterization of amino acid preference at membrane-water interfaces. Bioinformatics 18:608–616

Saitou N (2009) From selectionism to neutralism: paradigm shift of evolutionary studies. NTT Publishing Co., Tokyo. (In Japanese)

Tsuji T, Mitaku S (2004) Features of transmembrane helices useful for membrane protein prediction. Chem-Bio Informatics J 4:110–120

Part IV
Understanding Evolution Through the Phase Diagram of Life

Chapter 13
Definition of Species and Mysteries of Evolution

Keyword Definition of species · Species and genera · Gradualism · Punctuated equilibrium · Mega-extinction

We have already discussed that the nucleotide composition at each letter position of the codon is highly skewed from a completely random composition, which allows organisms to exist in a very stable manner. Furthermore, each species has a target nucleotide composition, and the nucleotide composition in individual organisms must be in the vicinity of that target. In other words, if you plot the fitness of an organism against the nucleotide composition from the entire genome sequence, it will be maximal at the composition of the target. Then, when mutations are introduced into the intracellular factors that determine nucleotide composition, the target's nucleotide composition jumps, and a new species is expected to be born. When we calculated the distance of nucleotide composition in the evolutionarily closest species pairs, we actually observed a jump in the phase diagram of life, the nucleotide composition space. When we calculated the distances for many genome sequences, the interspecies distances within a genus were clearly different from the intergenus distances. This leads to the hypothesis about evolution that the birth of a species is a random walk in the phase diagram of life. Furthermore, we discussed some of the mysteries of evolution on the basis of the phase diagram of life.

13.1 Fitness of Organisms in Nucleotide Composition Space

The stability of an organism is maintained by the harmony of many molecules. The concept of fitness can be used to understand the stability of the whole. However, it seems that no study has ever discussed the stability of an entire organism based on its genome sequence as a whole. In contrast, we already know that nucleotide composition at each codon letter position is an appropriate parameter to characterize the stability of an organism. Therefore, we will use nucleotide composition as a parameter to discuss the stability of the organism as a whole. Figure 13.1a shows the stability of an organism with nucleotide composition on the horizontal axis and fitness

S. Mitaku, R. Sawada, *Evolution Seen from the Phase Diagram of Life*, Evolutionary Studies, https://doi.org/10.1007/978-981-97-0060-8_13

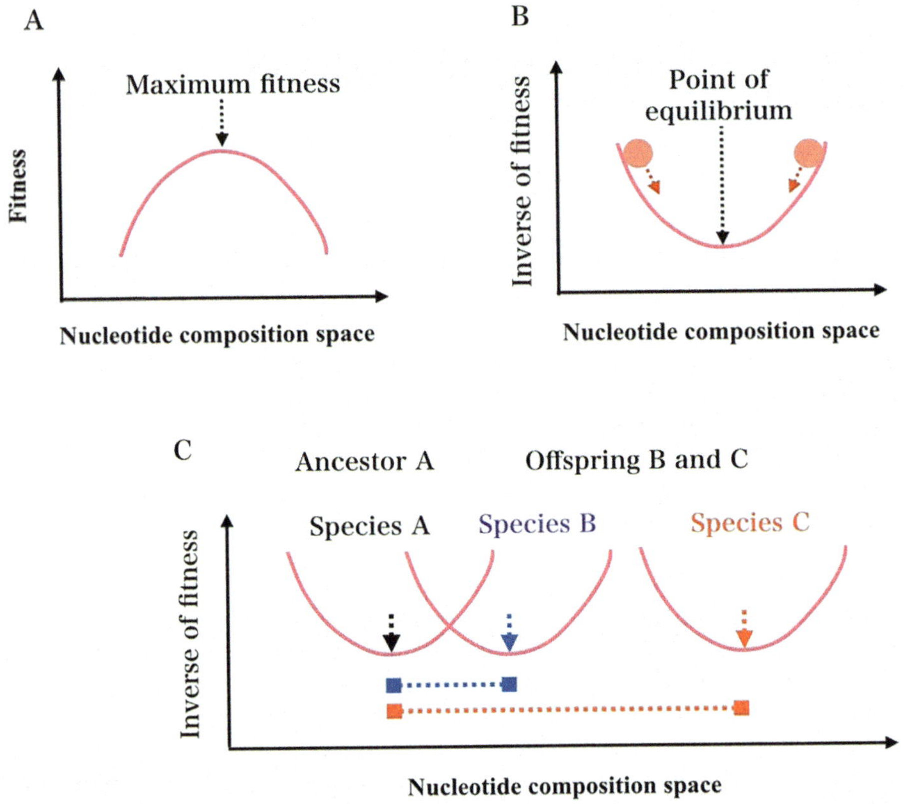

Fig. 13.1 The maximum fitness of each species at a given nucleotide composition is revealed in Part III (**a**). In physics, stable states are often defined by the minimum of an evaluation function. Therefore, using the inverse of fitness as the evaluation function, the most stable state of a species is represented by the minimum of the evaluation function (**b**). If the nucleotide composition of the genome sequence deviates from the minimum, there will be a restoring force in the direction of the minimum. The organism will then be extremely stable despite many mutations. It is natural that many mutations become neutral in the vicinity of the minimum. In this model, we can consider the change in nucleotide composition when two offspring species arise from an ancestral species. When ancestral species A diverged into offspring species B and C, the nucleotide compositions moved to different positions (**c**). Depending on how they move in compositional space, we can determine whether the change is gradual evolution or punctuated equilibrium

on the vertical axis. Even within the same species, there are differences in genome sequences among individuals. The spread of the fitness curve in this figure can be regarded as the frequency distribution of individuals at each nucleotide composition within the same species.

The maximum value of fitness represents the most average nucleotide composition as a species. The point here is that the target value of nucleotide composition is determined by intracellular factors, and fitness is maximal at that value. The further away the nucleotide composition is from the maximum value of fitness, the more unique the DNA sequence is as a species, and the further away from the average property. If it departs too far from the maximum, it loses stability as a species. Fig. 13.1b is just a graph with the inverse of fitness on the vertical axis, but it is just made in a way that is more acceptable to physicists. In physics, it is often expressed that the system becomes stable at the point where the evaluation function is minimized.

Now, in considering the evolution of organisms, we cannot avoid the question of what happens at the birth of a new species. To answer this question using actual whole-genome sequences, let us consider Fig. 13.1c, which shows the relationship between an ancestor and two offspring. Each species has a different target nucleotide composition. Thus, the two offspring will generally have different nucleotide compositions than the target ancestor species. The genome sequences of the ancestor and offspring cannot be directly compared. However, since offspring branch off, we can estimate what happens when a new species is born by examining the differences between the evolutionarily most closely related species. This is illustrated in a fitness diagram Fig. 13.1c.

Looking at the genome sequences of many species, it is likely that the nearest neighbor pairs are those that have recently branched off from the ancestral species. With such species pairs, we should be able to estimate how nucleotide composition changes when a new species is born. If there are large jumps in nucleotide composition targets when a new species is born, the genome sequence will be rapidly altered by many mutations. On the other hand, if there is not too large a jump, the genome sequence can change gradually. In other words, we may be able to distinguish between gradual evolution and punctuated equilibrium evolution.

13.2 How Does Nucleotide Composition Change When a New Species Is Born?

In this section, we will analyze the genome sequences of prokaryotes. There are two significant reasons to perform genome analysis on prokaryotes. First, prokaryotes have a much larger number of genome sequences. Another is that eukaryotes can define species by whether they reproduce sexually or not, but prokaryotes, which reproduce asexually, cannot define species as in that way. Therefore, defining species by genome sequence is especially useful for microbiologists.

Thus, the following analysis was performed on prokaryotes. First, sequence alignments were performed for 2664 species (2551 prokaryotes and 113 eukaryotes) to create a phylogenetic tree (Fig. 13.2). For this phylogenetic tree analysis, a neighbor-joining tree based on 16S rRNA gene sequences was constructed under the HKY distance matrix using PAUP* (version 4.0). For the pairs of species with the closest evolutionary distance, the distance in the eight-dimensional nucleotide composition space was calculated using Eq. (13.1).

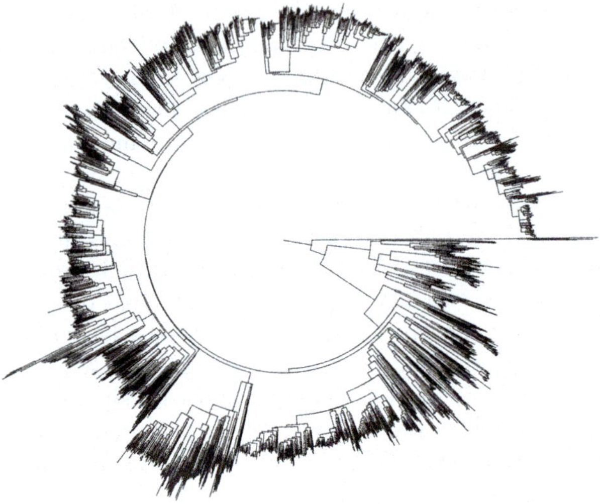

Fig. 13.2 Phylogenetic tree of 2551 prokaryotes (including 173 extreme environment organisms) and 113 eukaryotes. Prokaryotes are grouped into 853 genera

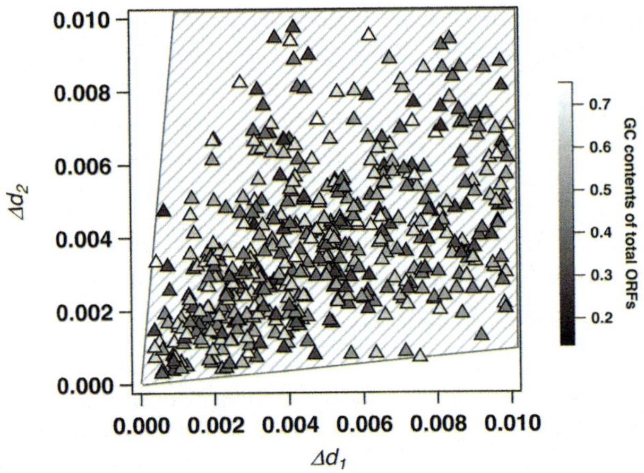

Fig. 13.3 Δd_1 vs. Δd_2 scatterplot shows the distance distribution between nearest neighbor species, in grayscale according to the GC content. As shown in Fig. 11.4, globally, the nucleotide composition of all species deviates significantly from a completely random composition. In contrast, this figure shows that locally at the birth of a new species, the changes are small and random

$$\Delta d_k = \sqrt{\sum_{i=A,T,G,C} \left[x_k^{\text{offspring1}}(i) - x_k^{\text{offspring2}}(i) \right]^2} \qquad (13.1)$$

where the subscript k represents the codon letter position 1 or 2. Additionally, x_k is the value of nucleotide composition, and Δd_k is the distance between the two nearest neighbor species. Figure 13.3 shows a scatter plot of the distance Δd_1 for the first letter vs. Δd_2 for the second letter. Triangles indicate the nearest pair of organisms in grayscale according to the GC content. The distances between the nucleotide compositions of the nearest species pairs are randomly scattered, as seen by various GC contents. The distances can be very close to zero or somewhat far apart. Additionally, there are no plots in the regions close to the horizontal or vertical axes.

Figure 13.3 is a zoomed-in graph of areas where distances are relatively close to zero. To obtain an accurate distance distribution for the nearest neighbor pairs, it is necessary to look at the areas without plots. However, some correction should be made in the areas where the distances are longer. The nearest neighbor pairs are the only genome pairs currently available to us. However, it is possible, especially if the distance is long, that there may be extinct species between the actual nearest neighbor pairs. The distance between nearest neighbor pairs should tend to include such extinct species in the pair.

$$\Delta D_{12} = \sqrt{\Delta d_1^{\,2} + \Delta d_2^{\,2}} \tag{13.2}$$

Therefore, we combined the distance between the first and second letters of the codon to obtain the overall distance according to Eq. (13.2). The result is shown in Fig. 13.4. Figure 13.4a shows three curves, one for the distance of the nearest pair and the other for the most remote pair in the same genus. These two curves tend to be nearly identical when the distance exceeds 0.04. Therefore, we hypothesized that the coincidence of these graphs is due to extinct species that would have been between the nearest pair. We then corrected the overall graph by assuming that the curve of the nearest pair and the curve of the remotest pair are the same at distances greater than 0.04. The corrected distance distribution is shown in Fig. 13.4b, and we found that the actual distance of the nearest pair is in the range of 0–0.01.

13.3 What Is Different Between Species and Genera When Viewed from the Composition Space?

Next, we examined how the distribution of distances between genera differs from that between species in the nucleotide composition space. The 2551 prokaryotic species belong to 853 genera. Therefore, we randomly selected one species from each genus among the 2551 species for analysis. The method of analysis was the same as for the interspecies case, and the distance of the nearest pair in the nucleotide composition space was calculated by Eqs. (13.1) and (13.2). Figure 13.5 shows the nearest genus pairs plotted as a scatter plot of Δd_1 vs. Δd_2.

Comparing Figs. 13.3 and 13.5, there are several similarities and significant differences. In the genus pair, as in the species pair, the points are plotted randomly regardless of GC content. In addition, they are similar in that there are no plots near both axes. The major difference, however, is that the nearest genus pair has no plot

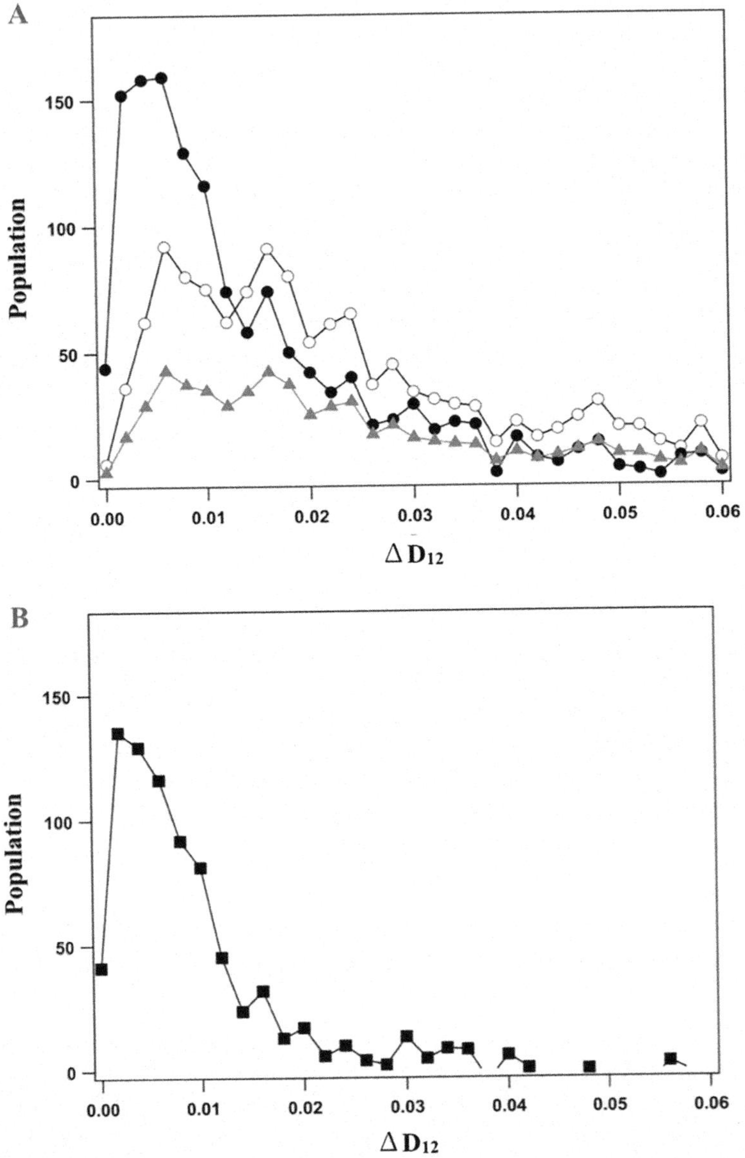

Fig. 13.4 (**a**) Histogram of distance distributions in nucleotide composition space at the birth of a new species. The distances between the evolutionarily closest and most distant species pairs within a genus are shown. If there is an extinct species between the closest species pair, it can be assumed that the distance distribution at long distances will be similar to the distance distribution of the furthest species pair. (**b**) Results corrected according to this assumption are shown. These results indicate that new species are distributed within a distance of approximately 0.01 in compositional space

in the region close to the origin. In this sense, the distribution of nearest genus pairs is quite different from that of species pairs.

Figure 13.6a is a histogram of the overall nearest neighbor genus distances calculated by Eq. (13.2). Figure 13.4b is also shown here for comparison of pairs of nearest neighbor genera and nearest neighbor species. Targets of nucleotide composition clearly show longer jumps between genera (circle mark) than between species (square mark). In addition, a graph of the number of accumulations is shown in Fig. 13.6b. It is clear that in the distribution of nearest-neighbor genus pairs, there are no plots in the region close to the origin. The cumulative number of nearest genus pairs begins to rise around a distance of 0.05. At a distance of 0.1, the number of accumulations of nearest-neighbor genus pairs is at most 10%, while the number of accumulations of nearest-neighbor species pairs is over 80%. From these results, we speculated that the difference between species and genera may lie in the size of the jump in compositional space. The jump in distance in compositional space may be an intrinsically important parameter for distinguishing between species and genera.

Figure 13.7 shows a model representation of the diversification of organisms during the evolutionary process. First, each species has its own unique position in the nucleotide space shown in Fig. 11.4. When a new species is created, its position jumps, but the size and direction of the jump are random. Furthermore, when the jump is small, a new species is born, and when the jump is large, a new genus is formed. The genome of an organism never deviates from the region shaded in

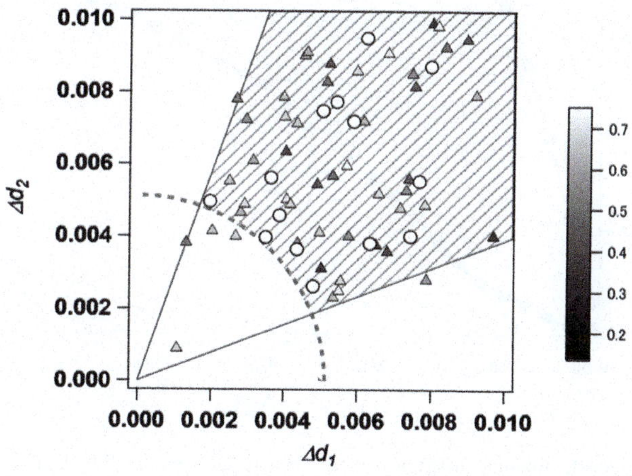

Fig. 13.5 One species was randomly selected from each of the 853 genera. The distance between the nearest neighboring genera was then calculated and plotted on a Δd_1 vs. Δd_2 scatter plot. The results showed that new genera move randomly within the compositional space. In the sense of being random, the birth of a genus and the birth of a species (Fig. 13.3) are similar. However, they also have completely different characteristics. Obviously, there is no plot near the origin. This means that when a new genus is born, it jumps considerably in compositional space. White circles are distances in ape pairs, very similar to plots in prokaryotes

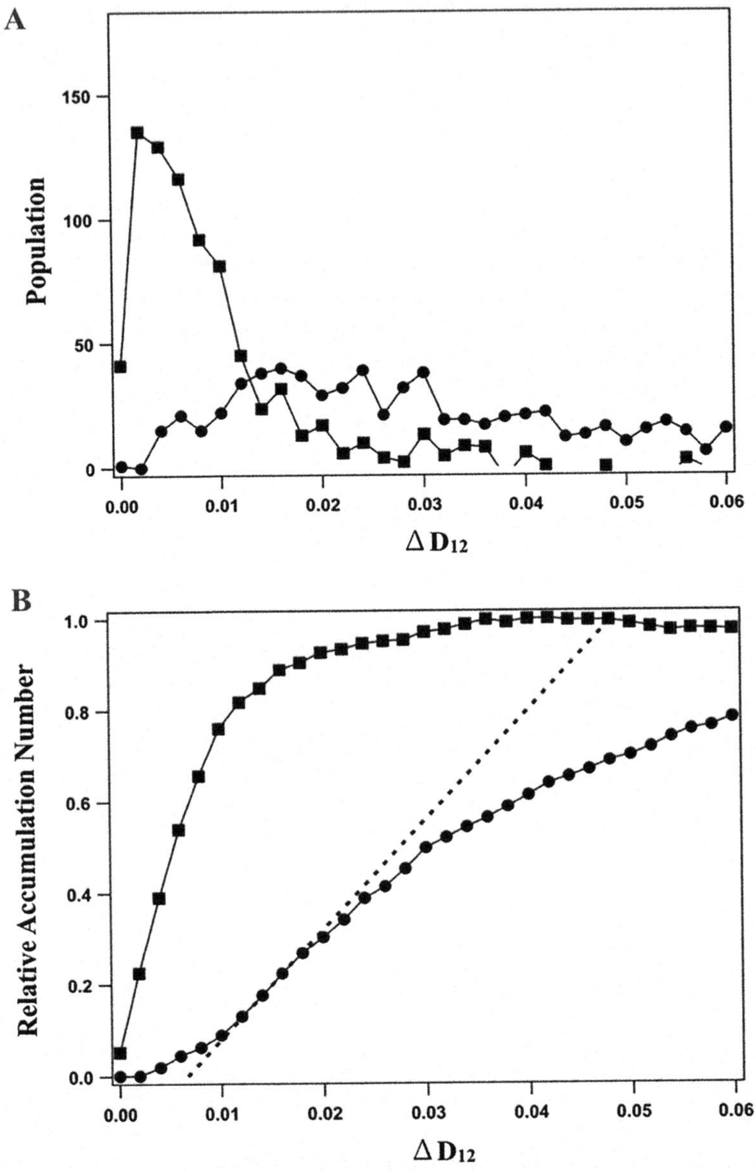

Fig. 13.6 (**a**) Overall distance histograms are shown to compare the distance distributions of nearest-neighbor genus pairs (circle mark) and nearest-neighbor species pairs (square mark) in prokaryotes. Clearly, the nearest neighbor genus pairs are distributed over longer distances. (**b**) Cumulative numbers are shown to facilitate understanding of this feature. Nearest-neighbor species pairs are distributed at distances shorter than approximately 0.01, while the nearest-neighbor genera are distributed at distances longer than 0.01. In other words, the distance distribution in compositional space is a useful parameter for distinguishing between species and genera

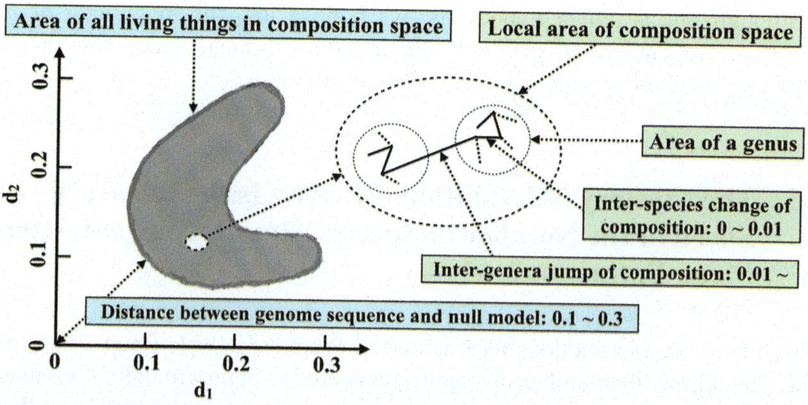

Fig. 13.7 Model diagram showing the characteristics of the distance distribution of species in nucleotide composition space. The global distance distribution, shown as a d_1 vs. d_2 scatterplot, is in the habitable zone, with the nucleotide composition of the genome sequence deviating significantly from a completely random composition. Locally, the species is on a random walk within the habitable zone. Furthermore, if the random walk steps are large enough, a new genus is created

Fig. 13.7. In other words, the evolutionary process can be thought of as a random walk within the phase diagram of life (nucleotide composition space).

13.4 The Mystery of Evolution (1): Does a New Species Emerge Gradually or Through Punctuated Equilibrium?

The question of whether new species emerge suddenly (punctuated equilibrium) or gradually (gradualism) is still open (Gould and Eldredge 1993; Zimmer and Douglas 2013). However, assuming mutations to intracellular factors that determine the phase diagram of life, the genome sequence makes it possible to discuss this issue as well.

For a new species to arise, many mutations must be introduced into the genome. Our view is that the introduction of many such mutations is caused by mutations in the intracellular factors that determine composition. That is, when a mutation occurs in an intracellular factor, it causes a jump in the nucleotide composition of the target. This causes a large number of mutations to be systematically introduced into the genome. If the jump is large enough, the mutation will rapidly enter the genome sequence, and the nature of the organism will suddenly change. Then, in the style of punctuated equilibrium, a new species will arise. In contrast, if the change in the target nucleotide composition is small, the mutation of the genome sequence will be gradual, and the birth of a new species will be gradual as well. In other words, both gradualism and punctuated equilibrium can occur in species evolution.

Although this is still a hypothesis, we believe that the magnitude of jumps between nearest-neighbor species pairs in compositional space can determine whether gradualism or punctuated equilibrium occurs.

13.5 The Mystery of Evolution (2): Why Is the Recovery Speed of the Number of Species Almost Constant After Mega-Extinction?

At least five mega-extinctions have occurred (Sepkoski 1981; Raup and Sepkoski 1982). Each time, the number of species recovered at approximately the same rate. This phenomenon is closely related to the mechanism of species birth. For the phenomenon of the birth of new species and genera, we have proposed the mechanism of random walks in the nucleotide composition space. We would then like to discuss the process of species recovery after mega-extinction based on the phase diagram of life.

As mentioned in the previous chapter, species-specific nucleotide composition is thought to be determined by intracellular factors (e.g., repair enzymes) (Fig. 8.1). Then, mutations are introduced into the DNA sequence of intracellular factors at a certain rate. The birth of new species will then occur at a constant rate. As a result, new species will always be created at a constant rate. The fact that the number of species is almost constant under normal conditions also means that the rate of birth of species and the rate of extinction of organisms are almost the same. In other words, the distribution of species is normally in a state of equilibrium, but after a sudden decrease in the total number of species due to a mega-extinction, a recovery process to a new equilibrium state should occur at a similar rate.

The five mega-extinctions we have seen thus far can be viewed as five experiments in which the number of species is rapidly reduced by nature. In each of these five experiments, the number of species recovered at approximately the same rate, and the experiments were replicated. The time constant for recovery is approximately ten million years. On the other hand, the actual number of species is estimated to be tens of millions (Trefil 1991). This means that, roughly speaking, several new species are born each year. It is usually assumed that extinctions occur at the same rate as births and that the number of species is in equilibrium. However, if the rate of extinction greatly exceeds this rate, the possibility of mass extinction arises. The current rate of extinction of organisms is estimated to be approximately 400 species per year, which is clearly an anomaly and could lead to a sixth mass extinction.

References

Gould SJ, Eldredge N (1993) Punctuated equilibrium comes of age. Nature 366:223–227

Raup DM, Sepkoski JJ Jr (1982) Mass extinctions in the marine fossil record. Science 215:1501–1503

Sepkoski J (1981) A factor analytic description of the Phanerozoic marine fossil record. Paleobiology 7:36–53

Trefil J (1991) 1001 things everyone should know about science. Doubleday, New York, NY

Zimmer C, Douglas JE (2013) Evolution: making sense of life. W.H. Freeman and Company, New York, NY

Chapter 14
Biological Hierarchies and Various Mutations in Genome Sequences

Keywords Hierarchies of organisms · Vertebrates · Eukaryotes · Multicellular organisms · Autocorrelation function

One of the most remarkable phenomena in biology is higher evolution of organisms such as domains and kingdoms. How the genome sequence changed during the birth of a new domain, kingdom, or phyla of an organism is one of the major questions. The purpose of this chapter is to estimate how the physical properties of proteins in the genome sequence relate to the higher taxa of organisms. We first focused on the distribution of charges, a long-range interaction. We converted all amino acid sequences from the genome sequence into charge sequences and calculated their autocorrelation functions. As a result, we observed a remarkable charge periodicity of 28 residues only in vertebrates. This indicates that vertebrates can be accurately classified from their genome sequences alone. Furthermore, an exclusive feature was observed between prokaryotes and eukaryotes in the charge distribution of all amino acid sequences. Namely, in the autocorrelation function of the charge sequences, a long-tailed positive correlation was observed only in eukaryotes. In general, we know that many types of mutations exist in the genome sequence and found that the insertion and duplication of long DNA sequences are closely related to the higher hierarchy of organisms.

14.1 Physical Analysis of Genome Sequences for Hierarchy of Organisms

There is a deep hierarchy of organisms, and these are classified according to various characteristics of the organisms. Since the genome sequence is the blueprint of an organism, it should be possible to classify the hierarchy of organisms based on the genome sequence alone. Based on this idea, we have been conducting research to clarify the relationship between genome sequences and the hierarchy of organisms. Our research policy is relatively simple and can be summarized in the following three points.

S. Mitaku, R. Sawada, *Evolution Seen from the Phase Diagram of Life*, Evolutionary Studies, https://doi.org/10.1007/978-981-97-0060-8_14

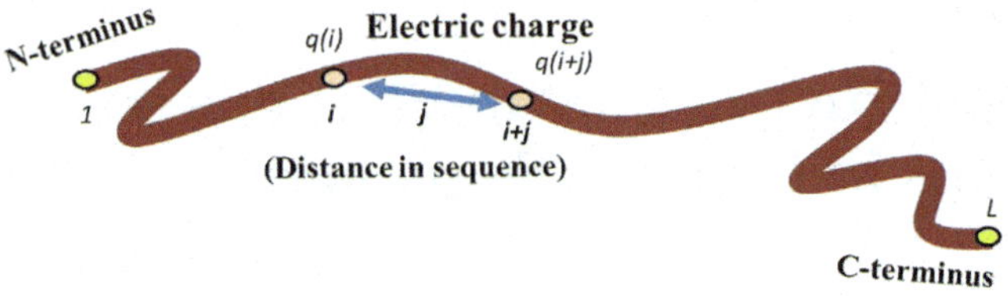

Fig. 14.1 Model diagram of the entire amino acid sequence in the coding region of the genome converted to a charge sequence. Here, we used the elementary charges (1, 0, −1) for each amino acid. From Mitaku (2015)

- As mentioned in the previous chapter, the nucleotide composition derived from genome sequences allows us to define species and genera. We also focus on the physical properties of the genome sequence for higher hierarchies of organisms.
- The simplest intensive parameter obtained from whole-genome sequences is the average value of the physical properties. The nucleotide composition used in the previous chapter is also an average. The next simplest possible parameter is periodicity or correlation around the mean value of the physical property, such as the autocorrelation function.
- While nucleotide and amino acid composition are important parameters, the physical properties of all amino acids derived from the genome sequence (hydrophobicity, charge, etc.) are also important properties.

Here, we focused on the autocorrelation function of the charge distribution of the total amino acid sequence (Fig. 14.1). We then investigated the correlation between the charge autocorrelation function and the higher-order hierarchy of organisms (Mitaku 2015).

The autocorrelation function is represented by Eq. 14.1.

$$\mathrm{AC}(j)=\frac{\sum_{k=1}^{N}\sum_{i=1}^{L(k)-j}\left[q(i)q(i+j)\right]}{\sum_{k=1}^{N}\left[L(k)-j\right]} \tag{14.1}$$

where q is the elementary charge (1, −1, 0), $L(k)$ is the length of the k-th amino acid sequence, and AC(j) represents the autocorrelation function at correlation length j.

14.2 Characteristics of Genome Sequences in Vertebrates

Various genome sequences were analyzed using Eq. 14.1 (Ke et al. 2008; Mitaku 2015) to obtain Fig. 14.2. Figure 14.2a shows the analysis of the fly genome as representative of the group of fungi, unicellular organisms, invertebrates, and plants. On the other hand, Fig. 14.2b shows the analysis results of the mouse genome as a representative of vertebrates. Although only representative genomes are shown here, all vertebrate genomes showed a 28-residue periodicity. Furthermore, the

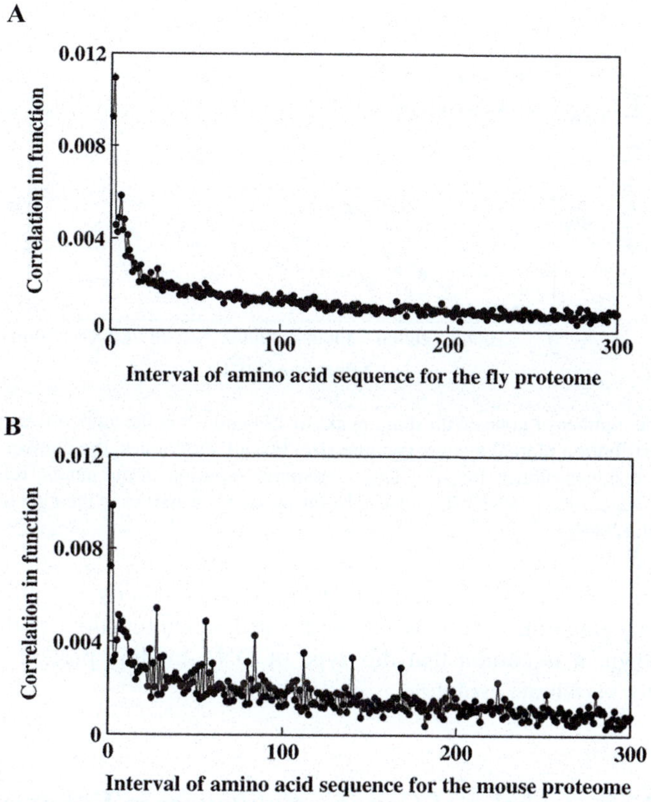

Fig. 14.2 (**a**) Autocorrelation function of the charge sequence of a fruit fly genome sequence representing invertebrates. (**b**) Autocorrelation function of the charge sequence of the mouse genome sequence representing vertebrates. All vertebrate genomes show very sharp peaks at multiples of 28 residues. On the other hand, no such sharp peaks were observed at residue 28 in non-vertebrate organisms. In other words, the presence of such sharp peaks is characteristic of vertebrate genomes. From Ke et al. (2008)

genomes of the remaining organisms did not show 28-residue periodicity. In other words, this feature was completely exclusive from a taxonomic point of view.

In an attempt to more clearly show the exclusivity between vertebrates and other organisms, we plotted the number of proteins showing 28-residue periodicity in Fig. 14.3. The horizontal axis is the genome size, and the vertical axis is the number of proteins with 28-residue charge periodicity. It is clear that the vertebrate group and the prokaryote, fungal, monera, invertebrate, and plant groups are on separate lines. The vertebrate group is clearly larger in number than the other organisms. The point where the two groups intersect is thought to be the genome size of organisms that evolved from prokaryotes to eukaryotes (Ke et al. 2007, 2008).

The fact that vertebrates and other groups of organisms are completely exclusive indicates that the property of 28-residue charge periodicity is very important for vertebrate evolution. Therefore, we performed a homology analysis of proteins that

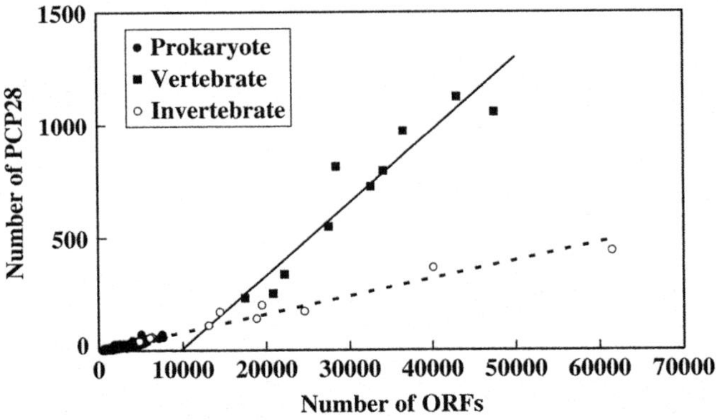

Fig. 14.3 The number of genes with sharp peaks at 28 residues in the autocorrelation function of the charge distribution plotted against genome size. For all vertebrates, the number of genes with a peak at 28 residues plotted linearly. This systematic variation in the autocorrelation function indicates that vertebrates can be distinguished by the physical properties of their genome sequences. From Ke et al. (2008)

exhibit 28-residue periodicity and found that this periodicity is similar to that of DNA-binding zinc finger proteins. Although the detailed functions of these proteins are not known, it is certain that this type of DNA-binding protein is extremely important for vertebrate evolution.

14.3 Characteristics of Genome Sequences in Eukaryotes

The charge autocorrelation function was also used in the comparison of prokaryotic and eukaryotic organisms. A plot of the autocorrelation function for the charge sequence of the *E. coli* genome sequence is shown in Fig. 14.4a. There is no correlation at all for many other prokaryotes. On the other hand, the autocorrelation function for the yeast genome shows a positive long-tail correlation (Fig. 14.4b). All other eukaryotic genomes also showed positive correlations with correlation lengths of a few dozen residues. In other words, prokaryotes and eukaryotes can be distinguished simply by analyzing the genome sequence (Ke et al. 2007, 2008).

The reason why the charge distribution of genome sequences clearly differs between prokaryotes and eukaryotes may be due to differences in cell structure. The most important feature of eukaryotes is that they have a nucleus. DNA, which carries genetic information, is located in the nucleus, but protein synthesis takes place in the cytoplasm outside the nucleus. Therefore, after proteins are synthesized in the cytoplasm, they must be selectively transported through the nuclear membrane into the nucleus. The proteins responsible for this process are called importins. The signal sequences that bind to importins are called nuclear localization signals (NLSs) and are known to contain many positive charges. However, there is no definitive

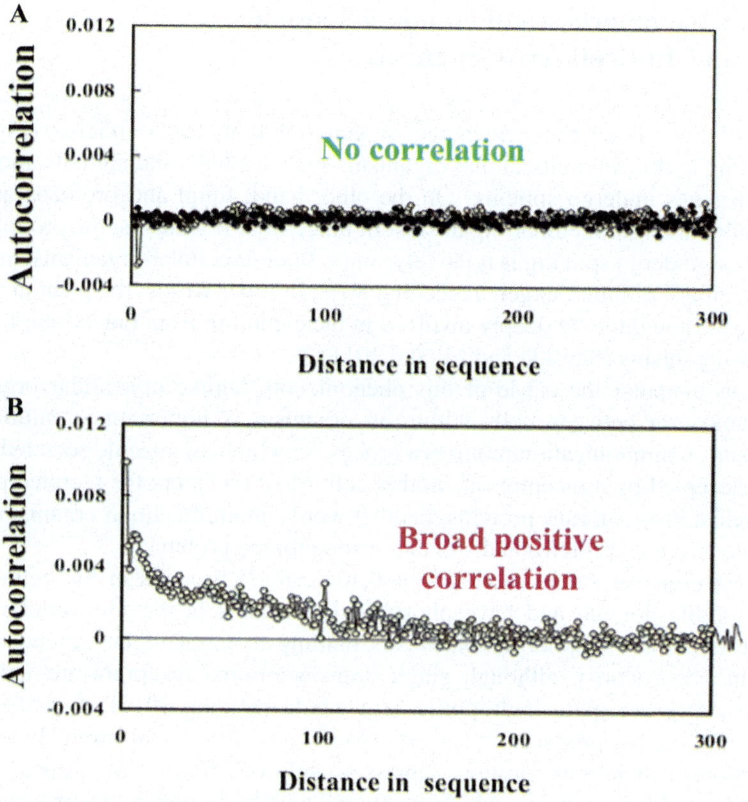

Fig. 14.4 Autocorrelation functions for eukaryotic and prokaryotic charge distributions. (**a**) Autocorrelation function of the *E. coli* genome as a representative of prokaryotes. (**b**) Autocorrelation function of the yeast genome as a representative of eukaryotes. All prokaryotic genomes were completely uncorrelated. On the other hand, all eukaryotic genomes showed positive long-tail correlations. This positive long-tail charge correlation is a physical feature of eukaryotic genomes. Adapted in part from Ke et al. (2008)

NLS sequence, and some degree of ambiguity seems to be tolerated. Thus, the long-tail correlation of charge distribution in eukaryotes is presumably due to signal peptides such as NLS sequences. Using physical parameters derived from genome sequences, eukaryotes and prokaryotes can be exclusively discriminated, and a phase diagram between them can be drawn.

14.4 Characteristics of Genome Sequences in Multicellular Organisms

Chapter 9 discussed the percentage of genes that undergo splicing (Fig. 9.7). According to this, in multicellular organisms such as plants and animals, more than 80% of genes undergo splicing. On the other hand, fungi and protozoa are both unicellular and multicellular organisms. In unicellular organisms, the proportion of genes that undergo splicing is relatively small. In multicellular organisms, however, the percentage is much larger, exceeding 80%. In other words, the phenomenon of splicing is thought to be deeply involved in the evolution from unicellular to multicellular organisms (Sawada and Mitaku 2011).

Let us consider the cause of this phenomenon. Unlike unicellular organisms, communication between cells within an organism is important in multicellular organisms. Communication requires a process in which a molecule secreted by one cell is accepted by a receptor on another cell. Most receptors for secreted proteins are single transmembrane proteins. In other words, in multicellular organisms, there are many secreted proteins and single transmembrane proteins.

Single transmembrane receptors bind to secreted proteins in the extracellular water-soluble domain and have phosphorylation sites in the intracellular water-soluble domain. Phosphorylation allows binding to intracellular second messengers. In other words, although single transmembrane receptors are membrane proteins, they are quite hydrophilic as a whole (Mitaku 2015). Nonetheless, as shown in Chap. 9, prokaryotic and eukaryotic genomes contain roughly the same proportion of membrane proteins, approximately one-fourth. We suspect that the fact that membrane proteins, which are overall very hydrophilic, can be formed may be closely related to the development of the splicing process. Furthermore, as mentioned in Chap. 9, it seems that the process of splicing is also made by random processes. In this sense, we can be sure that some physical process is involved in the evolution to multicellular organisms.

14.5 The Relationship Between the Hierarchical Structure of Organisms and the Physical Properties of Genome Sequences

It is usual to approach the whole-genome sequence by clarifying the function of many genes. In contrast to this bottom-up approach, we took a top-down approach focusing on the physical regularity of the entire genome sequence or all amino acid sequences. We then discussed the physical properties in the hierarchy of organisms and clarified the phase diagram of life.

First, we showed that the eight-dimensional space consisting of the nucleotide composition of the first and second letters of the codon corresponds to the phase diagram of life. We also showed that species and genera can be defined by the eight nucleotide compositions as parameters. Furthermore, we were able to distinguish vertebrates exclusively from other organisms by the charge periodicity of 28 amino acid residues. Positive correlations in the autocorrelation function of charge sequences also allowed us to exclusively distinguish eukaryotes from prokaryotes. Multicellular organisms are characterized by splicing, and there is evidence that this too is formed by random processes. In other words, each level of the organism reflects a different physical regularity of the genome sequence.

As shown in Fig. 7.4, the amino acids that perform the function of a protein are only a small fraction of the total amino acids. In other words, the system is composed of the physical regularities of the entire genome, and the addition of a small number of amino acids is thought to add functionality to the system. In the first half of the final chapter, we will discuss what viruses are in terms of the phase diagram of life. Finally, we will consider the physics of the genome sequence of organisms and discuss the remaining questions.

References

Ke R, Sakiyama N, Sawada R, Sonoyama M, Mitaku S (2007) Human genome includes many proteins with charge periodicity of 28 residues. Jpn J Appl Phys 46:6083–6086

Ke R, Sakiyama N, Sawada R, Sonoyama M, Mitaku S (2008) Vertebrate genomes code excess proteins with charge periodicity of 28 residues. J Biochem 143:661–665

Mitaku S (2015) A modern approach to biological science. Kyoritsu-Shuppan Co., Tokyo. (In Japanese)

Sawada R, Mitaku S (2011) How are exons encoding transmembrane sequences distributed in the exon–intron structure of genes? Genes Cells 16:115–121

Chapter 15
Analysis of Viruses, and the Fusion of Biology and Physics Through the Phase Diagram of Life

Keywords Phase diagram of life · COVID-19 · Zoonotic viruses · Extraterrestrial life · Bioethics

The study of genome sequences by the phase diagram of life is still in its early stages. One of the major remaining questions is whether viruses are living organisms in terms of their genome sequences. Both viruses and organisms have genome sequences. However, because they have very different life cycles, we do not know if the genome of a virus is in the same region of nucleotide composition space as that of a group of organisms. Therefore, we analyzed where the genome sequence of COVID-19 is located in the nucleotide composition space and how it migrated after becoming zoonotic. As a result, we first found that the viral genome is located in the phase diagram of life. After the virus became zoonotic, the genome sequence moved unilaterally within the nucleotide composition space. This change can apparently be explained by the fact that when the virus has changed hosts, the targets of nucleotide composition have changed.

This book is only the beginning of this field, and there are still many remaining questions. At the end of this book, we summarize these issues.

15.1 Virus Genome in Nucleotide Composition Space

When we started writing this book, we did not intend to address the virus issue. However, the COVID-19 pandemic occurred at the end of 2019 (Zhou et al. 2020). COVID-19 has continued to have a significant impact on society for more than 3 years. Therefore, in the last chapter of this book, we decided to analyze the viral genome through the phase diagram of life.

Viruses and organisms are the same in that they both have genomes. However, viruses are always dependent on the host's genome processing system and cannot produce the next generation of viruses using only their own genome. Thus, we do not know if the viral genome will plot in the region of the phase diagram of life in Fig. 11.4. On the other hand, if the nucleotide composition of the genome is

S. Mitaku, R. Sawada, *Evolution Seen from the Phase Diagram of Life*, Evolutionary Studies, https://doi.org/10.1007/978-981-97-0060-8_15

determined by the host organism's genome processing system (intracellular factors), then the nucleotide composition of the viral genome should also be within the phase diagram of life. In other words, whether the viral genome is within the phase diagram of life determines whether the nucleotide composition is determined by intracellular factors. Therefore, we first examined the location of the viral genome in Fig. 11.4. We also examined how the nucleotide composition of SARS-CoV-2 changed over time after zoonotic infection. For this analysis, we used a database of data obtained primarily in Japan (Shu and McCauley 2017).

In Fig. 15.1, the eukaryotic genome (gray), mammalian genome (blue), human genome (green), and SARS-CoV-2 virus genome (red circle) are plotted on a two-dimensional scatter plot of d_1 vs. d_2. This is the same figure as Fig. 11.4, but prokaryotic data not relevant to the discussion here have been removed. In this figure, the SARS-CoV-2 genome is located near the genome of its host, humans or other mammals. Viruses and organisms have very different life cycles and are therefore in very different environments. Nevertheless, the fact that their nucleotide compositions are similar strongly suggests that nucleotide composition is not affected by the environment but is determined by the host's genome processing system (intracellular factors).

However, the SARS-CoV-2 genome is slightly displaced from the human genome, which may be due to the structure of the SARS-CoV-2 genome. The virus has a membrane on its surface, and the proportion of membrane proteins (approximately 42%) is much larger than the proportion of membrane proteins in normal

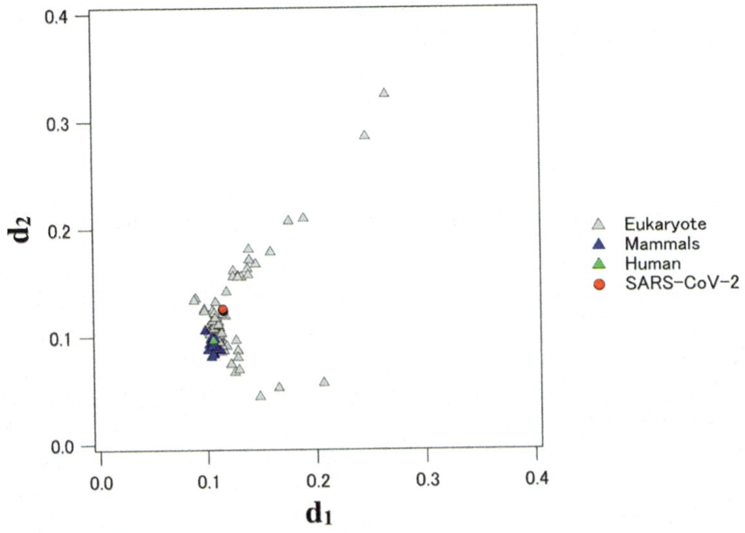

Fig. 15.1 Nucleotide compositions of eukaryotes (gray triangles), mammals (blue triangles), humans (green triangles), and SARS-CoV-2 (red circles) are plotted on a d_1 vs. d_2 scatter plot. Clearly, SARS-CoV-2 is in a region close to mammals, indicating that the virus is the same as the organism in terms of the phase diagram of life

organisms (approximately 1/4). The percentage of membrane proteins is related to the second letter of the codon. The more membrane proteins there are, the more their position in the nucleotide composition space is shifted from the host genome. For this reason, the SARS-CoV-2 genome is considered to have slightly larger d_1 and d_2 than the human genome.

Now, another question is which animal coronavirus first infected humans and how the coronavirus genome has changed since then (Temmam et al. 2022). However, since we do not know enough about this issue, we decided to use the genome sequence of the RmYN02 virus studied in Yunnan, China, in 2019 as a comparison. This is because the RmYN02 virus was analyzed near the time and place where the first COVID-19 patients appeared (Zhou et al. 2020). On the other hand, we did not use genomic data after 2020, when COVID-19 was already a zoonosis. We were interested in examining the direction of movement in the nucleotide composition space at and after the time the coronavirus became zoonotic.

Since 2019, the genome of the human virus SARS-CoV-2 has been rapidly changing, and the changes in the number of patients infected have interesting features. One feature is that there have been several waves of infection. The number of patients increases rapidly and then declines after approximately 3–4 months. Although there is a vaccination effect against the variants, this feature does not seem to be solely attributable to the vaccine. In other words, the waves of viral infections seem to have a certain lifespan. Another feature is that these waves of infection seem to settle down after a few years. We examined the changes in the viral genome to determine whether these phenomena can be explained by the phase diagram of life.

Figure 15.2 is a scatterplot of d_1 vs. d_2 for three bat species, humans, and the corresponding coronaviruses. Figure 15.3 is a magnified view of the viral portion of Fig. 15.2. Bat virus (light blue) is RmYN02 data analyzed in China in 2019. Various time periods are also color-coded for human SARS-CoV-2 viruses (Japan SARS-CoV-2 Database n.d.). The position of SARS-CoV-2 in the nucleotide composition space is very scattered but shows a gradual upward shift over time. Figures 15.2 and 15.3 show blue vectors. These are vectors in the direction from the host bat genome to the human genome. It should be noted that the direction of movement of the host genome and that of the viral genome are generally similar, although slightly offset.

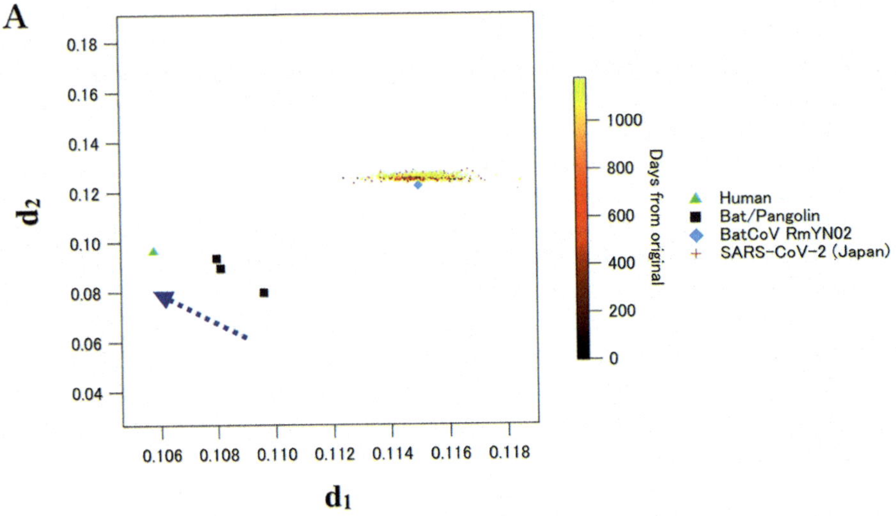

Fig. 15.2 Coronavirus genome and host genome plotted in compositional space: for the host, three bat species (black squares) and humans (green triangles); for the virus, bat coronavirus RmYN02 (light blue diamond) from China (2019) and human coronavirus SARS-CoV-2 (crosses). For SARS-CoV-2, the time course of the genome is plotted in colors. Vectors from the bat genome to the human genome are indicated by blue arrows

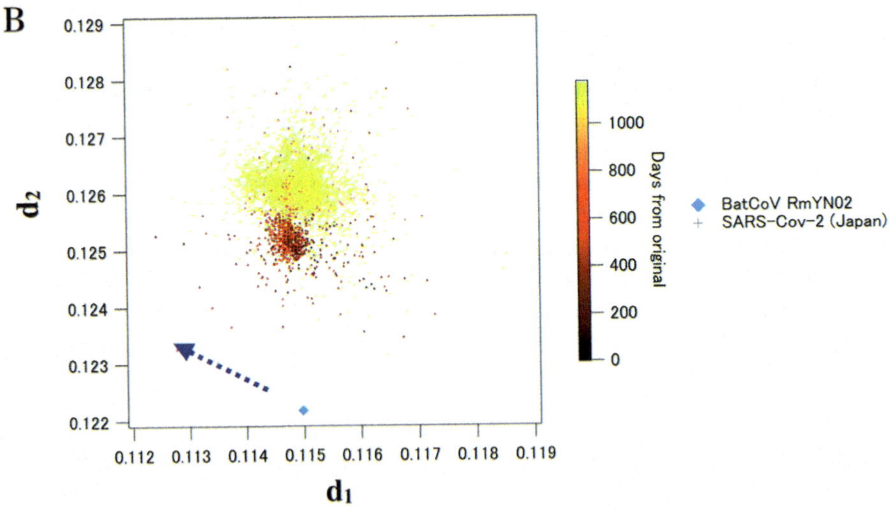

Fig. 15.3 Enlarged view of the virus portion of Fig. 15.2. Vectors about the host are also shown, but the direction of movement of the viral genome is slightly off by approximately 60 °. However, it is noteworthy that they are generally moving in the same direction and continue to move after infection

15.2 What Happened to the Genome Sequence After the Coronavirus Became a Zoonotic Disease

In general, genome sequence variation can be viewed as a random walk in the nucleotide composition space (Chap. 13). Thus, the large variation in the data for SARS-CoV-2 in Fig. 15.3 is reasonable if understood as a random walk. In general, random changes are smaller when the data are averaged. Unidirectional changes, on the other hand, remain the same when averaged. This can be used to distinguish between random and unidirectional changes. Therefore, we took the monthly average of the data in Fig. 15.3 to obtain Fig. 15.4.

The monthly changes in the graph are compact, and it is clear that the large spread in the data in Fig. 15.3 is due to a random walk. However, it is also clear that there is a unidirectional change (mainly an increase in d_2) that cannot be understood by random walk alone. Therefore, the unidirectional component of SARS-CoV-2 genomic variation requires scientific consideration.

Figure 15.5 is a graph of time variation by month. Figures 15.5a and b show changes in d_1 and d_2, respectively. The distance d_1 in the compositional space at the first letter of the codon is almost a random change, whereas the distance d_2 at the second letter of the codon shows a unidirectional change with random changes. After 2022, both d_1 and d_2 are almost constant.

To understand these phenomena, we use the concept of fitness, illustrated in Fig. 13.1. To discuss the stability of the genome sequence of a single species, it is sufficient to consider the fitness curve of that species. However, this is not sufficient to

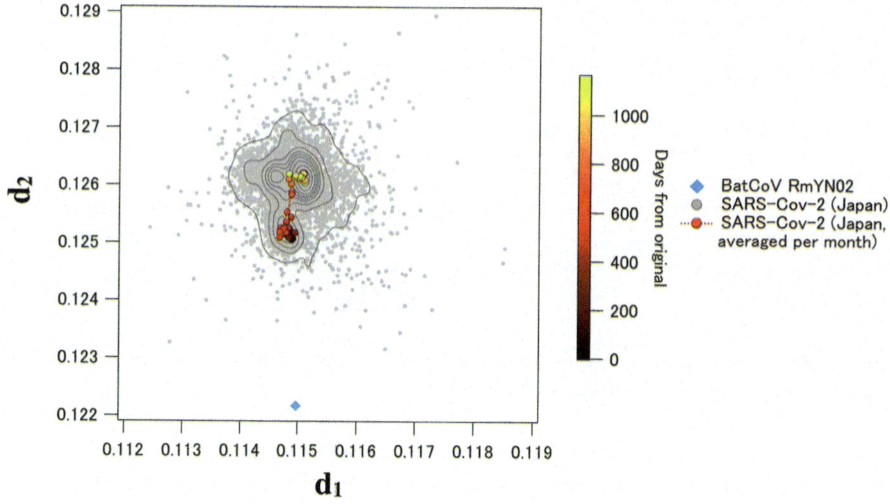

Fig. 15.4 Monthly average nucleotide composition calculated from the genome sequence of the SARS-CoV-2 virus. The gray markings are plots for each mutant type, and the color coding from brown to yellow represents the change in monthly average nucleotide composition. It is clear that the variants are moving in one direction. The contour lines indicate the density of viral variants

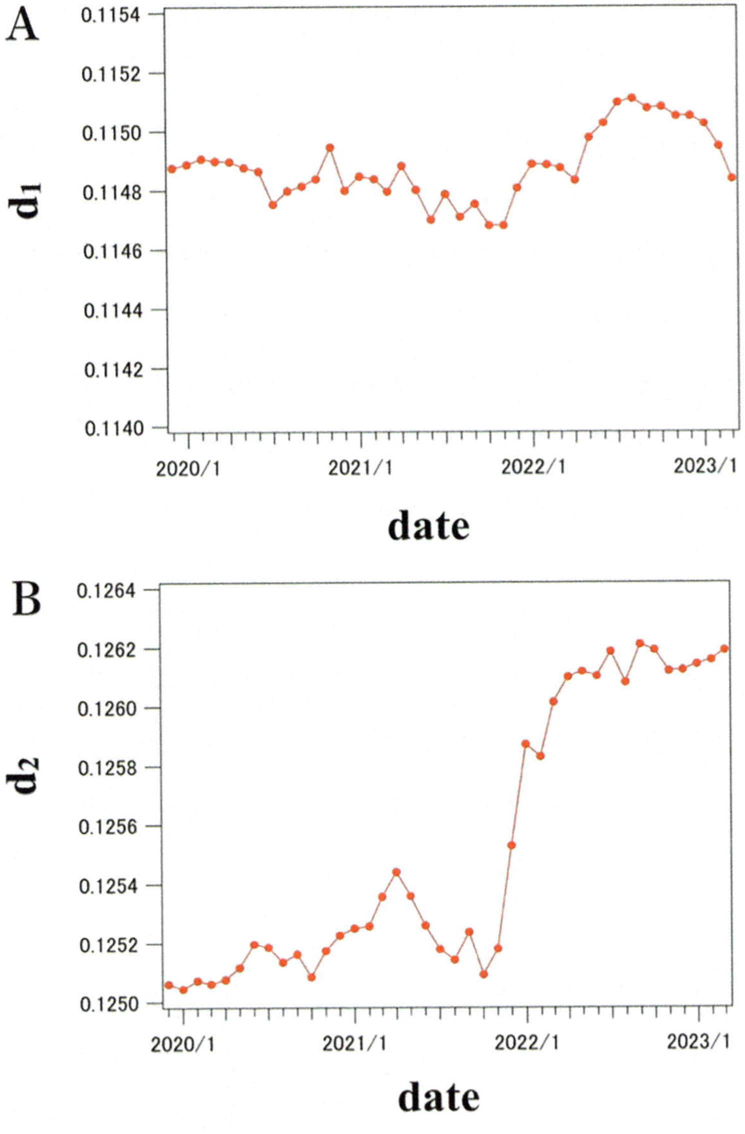

Fig. 15.5 The average monthly time change for d_1 is nearly constant (**a**), while the average monthly time change for d_2 increases unilaterally (**b**)

discuss the behavior of a viral genome when it changes hosts. When a virus changes hosts, one must consider the relationship between the fitness curves of the two hosts.

The two species will each exhibit a fitness maximum at a position unique to each species in nucleotide composition space. We cannot know about the spread of the fitness curves for each species. However, the fact that the virus has changed host organisms indicates that the fitness curves are close enough for the viral genome to

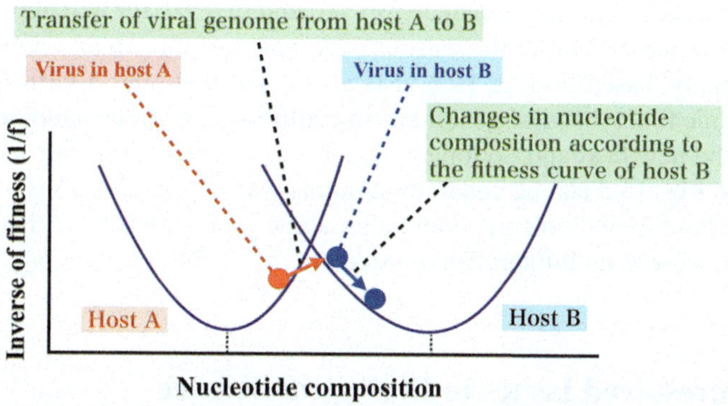

Fig. 15.6 Schematic diagram of the inverse fitness curves of host A (bat) and host B (human) in nucleotide composition space and the migration of the coronavirus genome. When the fitness curves of two hosts overlap, the viral genome can change hosts. When the host changes, the viral genome moves toward the new host's fitness curve. The viral genome then moves toward the optimal nucleotide composition of the new host's fitness curve

move across the two hosts. In fact, since the coronavirus (SARS-CoV-2) became infectious to humans in 2019, as shown in Fig. 15.6, the coronavirus genome has jumped from the fitness curve of the previous host (probably bats) to the fitness curve of the later host (humans).

Since the overall nucleotide composition of the bat and human genomes is actually close, we can assume that the fitness curves overlap. When a virus infects a host, the viral genome can be considered a subsequence of the host genome sequence. Such subsequences will be distributed around the target composition of the whole genome in compositional space. The percentage of membrane proteins in coronaviruses is approximately 42%, which deviates considerably from approximately 1/4 of the host membrane proteins. Thus, the nucleotide composition of the coronavirus will deviate somewhat from the target composition of the host. The plot in Fig. 15.2 actually shows a slight bias in the nucleotide composition of the host and the virus.

When the host of the virus changes from bats to humans, the genome processing system also switches from bats to humans. As a result, the composition of the target also makes the jump from bats to humans. However, at the time of the jump, the nucleotide composition of the viral genome should be biased away from its most stable values in humans. Then, by mutation, the viral genome moves unidirectionally toward the most stable value. The unidirectional movement of SARS-CoV-2 seen in Fig. 15.4 can be understood in this way. Indeed, the fact that the bat-to-human vector in nucleotide composition space is nearly identical to the direction of the vector in the viral genome (Fig. 15.2) is compelling evidence for this idea.

Let us now consider why the wave occurs in infected individuals. From the discussion in Fig. 15.6, when the host of a virus changes, the nucleotide composition inevitably changes little by little in one direction. As a result, some variants will not maintain the same genomic sequence forever, and one variant will appear to have a lifespan. This may be one of the reasons why waves of infections occur.

Additionally, based on Fig. 15.6, it is also noted that after a virus has changed hosts for some time, the viral genome is in equilibrium with the genome processing system of the new host and no longer changes in one direction. The recent plateau over time in Fig. 15.5 can be understood in this way. If so, the COVID-19 mutation would only have a random component after a few years. We believe that COVID-19 would then become an influenza-like infection.

15.3 Unresolved Issues in Biological Science from the Perspective of the Phase Diagram of Life

We have discussed biology and evolution using the new concept of the phase diagram of life. At the end of this book, we consider some issues that should be studied in the future with respect to the concept of the phase diagram of life.

- At the beginning of this chapter, we showed that the genome of zoonotic SARS-CoV-2 is very close to the human genome in nucleotide composition space. Furthermore, there are many other zoonotic viruses besides SARS-CoV-2. We believe that the genomes of these zoonotic viruses are all in close proximity to the human genome in nucleotide composition space. However, this must be proven by actual analysis. Furthermore, some viruses have membranes, and some do not. They should have systematically different protein distributions, and the former should clearly have a higher proportion of membrane proteins. Then, the genomes of the two groups of viruses are likely to be in different positions in the nucleotide composition space. These predictions can be quickly confirmed by examining the many viral genomes already available.
- Astrobiologists believe that comparing terrestrial and extraterrestrial life can provide information about the definition of life (Smith 2016). Following this idea, the search for extraterrestrial life is underway. It is believed that terrestrial life originates from organic matter in water, and the presence of water and organic matter is considered a prerequisite for life (Cockell 2018). However, assuming that extraterrestrial and terrestrial lives are made of the same material, a comparison based on different criteria is also necessary. In other words, assuming that extraterrestrial life also has a genome, it is a very interesting question whether it is in the same phase diagram of life as terrestrial life.
- In Chap. 11, we stated that a simple random mutation will result in a constant distribution of protein types if the nucleotide composition is highly biased. We then showed that, in fact, random mutations with a biased nucleotide composition will result in a constant ratio of membrane to soluble proteins in each genome. However, there are various types of soluble proteins, and it has not yet been confirmed whether their distribution is constant. Or, just as the number of transmembrane

helices correlates with function, the fold of a soluble protein may also correlate with function. The concept of protein universes was discussed as a suggestion that there is order in the distribution of protein 3D structures (Koonin et al. 2002). However, the distribution of protein 3D structures at the genome level has not yet been investigated. Recently, a highly accurate protein 3D structure prediction system was developed (Jumper et al. 2021). It is hoped that such a study will clarify the distribution of protein 3D structures throughout the genome.

- There is a deep hierarchy in the classification of organisms. In Chaps. 13 and 14, we discussed the classification of species and genera, prokaryotes and eukaryotes, and invertebrates and vertebrates and showed that the physical properties of the genome sequence are deeply involved in these classifications. However, even deeper classifications include domain, kingdom, phylum, order, family, genus, and species. Our research suggests that the physical properties of the genome sequence are involved in the classification of these organisms. There are indeed many taxonomic categories of organisms, but we have studied only a small fraction of them. We predict that some physical property of the genome may be involved in each of these other biological classifications.
- We analyzed the whole-genome sequences of many species and found that their nucleotide compositions were in regions that deviated significantly from a completely random composition. We then hypothesized that the bias in nucleotide composition was controlled by intracellular factors, possibly due to error bias in the repair system. We also stated (Chaps. 13 and 14) that if the bias in nucleotide composition was due to intracellular factors, it should be able to explain some of the mysteries of evolution. However, the identity of the intracellular factor must ultimately be confirmed experimentally. That is, we must prove experimentally that bias in the probability of nucleotide generation by intracellular factors does indeed occur. This is the greatest challenge remaining for our research.
- Finally, we must discuss bioethics. We have reasoned that the nucleotide composition space, the phase diagram of life, is determined by intracellular factors. In other words, the diversification of organisms can be attributed to mutations to intracellular factors. This implies that genome editing technology can be used to artificially create new species. However, from two perspectives, such research raises major ethical issues. One is the issue of ecological collapse. In nature, the establishment of a new species depends on a process of natural selection by the surrounding ecosystem. The new species may disappear quickly. However, there is also a great possibility that it will have some effect on many existing organisms. Therefore, such experiments must be conducted with great care. Another problem is the attempt to mutate intracellular factors in primates. In nature, humans also originated from primate ancestors. However, we believe that any experiment that might produce another organism that resembles humans should be prohibited.

We believe that genome research based on the "phase diagram of life" has the potential to develop in many directions in the future. We hope that the concept of the "phase diagram of life" will lead to the fusion of biology and physics and the birth of a new academic field.

References

Cockell CS (2018) The equations of life. Atlantic Books, London

Japan SARS-CoV-2 Database (n.d.) Phylogenetic analysis of SARS-CoV-2 diversity in Japan. https://nextstrain.org/community/kkosaki/ncov/japan

Jumper J, Evans R, Pritzel A et al (2021) High accurate protein structure prediction with AlphaFold. Nature 596:583–589

Koonin EV, Wolf YI, Karev GP (2002) The structure of the protein universe and genome evolution. Nature 420:218–223

Shu Y, McCauley J (2017) GISAID: Global initiative on sharing all influenza data—from vision to reality. Euro Surveill 22(13):30494. https://doi.org/10.2807/1560-7917.ES.2017.22.13.30494; https://gisaid.org; https://www.epicov.org/

Smith K (2016) Life is hard: countering definitional pessimism concerning the definition of life. Int J Astrobiol 15:277–289

Temmam S et al (2022) Bat coronaviruses related to SARS-CoV-2 and infectious for human cells. Nature 604:330–336. https://doi.org/10.1038/s41586-022-04532-4

Zhou H et al (2020) A novel bat coronavirus closely related to SARS-CoV-2 contains natural insertions at the S1/S2 cleavage site of the spike protein. Curr Biol 30:2196–2203

The manufacturer's authorised representative in the EU is Springer
Nature Customer Service Centre GmbH, Europaplatz 3, 69115 Heidelberg,
Germany. If you have any concerns regarding our products, please
contact ProductSafety@springernature.com

Printed and bound by CPI Group (UK) Ltd, Croydon, CR0 4YY
01/10/2026
02225883-0009